INTRODUCTION TO
PHYSIOLOGICAL PLANT ECOLOGY

Introduction to Physiological Plant Ecology

P. BANNISTER

BSc, PhD
Department of Biology
University of Stirling

SECOND PRINTING

BLACKWELL SCIENTIFIC PUBLICATIONS
OXFORD LONDON EDINBURGH MELBOURNE

© 1976 Blackwell Scientific Publications
Osney Mead, Oxford
8 John Street, London WC1
9 Forrest Road, Edinburgh
P.O. Box 9, North Balwyn, Victoria, Australia

First published 1976
Reprinted 1978

British Library Cataloguing in Publication Data

Bannister, Peter
    Introduction to physiological plant ecology.
    Bibl.—Index
    ISBN 0–632–08980–6
    1. Title
    581'.1 QK711.2
    Plant physiology
    Botany—Ecology

Distributed in the
United States of America by
Halsted Press, a division of
John Wiley & Sons Inc, New York
and in Canada by
J. B. Lippincott Company of
Canada Ltd, Toronto

Set in Great Britain by
Enset Ltd., Midsomer Norton, Bath
Printed in Great Britain by offset lithography by
Billing & Sons Ltd, Guildford, London and Worcester

Contents

Preface

In writing this book, I have tried to relate the ecology of plants and their physiological responses to their environment. I have presupposed that the readers of this book have an elementary knowledge of general botany, plant ecology and plant physiology but chapters on the aerial and soil environments have been included in order to give the reader a simple introduction to these areas. Throughout the book reference is made to more specialized texts and articles so that areas which have necessarily been treated superficially can be studied in greater detail if so desired. The references cannot be comprehensive: many examples are drawn from British and European sources, not through chauvinism but because they reflect my own experience. Consequently, I hope that this book will complement such American texts as Daubenmire's *Plants and Environment* (1974) and also provide an introduction to continental, particularly German, physiological ecology which has not only provided one of the original treatises on physiological ecology, Schimper's *Pflanzegeographie auf physiologischer Grundlage* (1898), but also some of the most recent, Larcher's *Ökologie der Pflanzen* (1973) and Kreeb's *Ökophysiologie der Pflanzen* (1974). The book is designed primarily as an undergraduate text and is based on courses that I have given at Stirling University. However, it should also prove useful to schoolteachers and senior pupils in schools, particularly as contemporary syllabuses and project work often emphasize the interactions of organisms and environment, and I hope that the professional ecologist will find the book of interest, both in itself and because of its reference to more specialized literature.

The organization of the book is influenced by its subject matter. Plants interact with both the aerial and soil environments and consequently a consideration of the aerial environment is followed by chapters on plant responses to this environment; a description of the soil environment then precedes a chapter on water relations that links the aerial and soil environments; and a chapter on mineral nutrition is wholly concerned with the soil. The last chapter deals with other interactions, particularly with biological and man-made environments. The book tries not to consider the influence of each environmental factor in turn as such an approach can lead to too great an emphasis on the environment and makes it more difficult to discuss the complex interactions between a plant function and a number of environmental influences. Instead, this book has tried to emphasize physiological functions, such as photosynthesis and respiration, water relations and mineral nutrition and place them in an environmental context. Environmental influences have therefore been considered at various stages in development such as at germination and establishment and in the mature plant, during the dormant and growing seasons, and upon critical functions such as stomatal movement which can influence photosynthesis, water relations and energy balance. Plant responses to environmental factors are considered both in terms of the tolerance of extreme levels (i.e. insufficiency and superabundance) and the reaction to the 'normal' range.

Some mention must also be made of aspects that are not dealt with in this book. There is no detailed consideration of specific experimental techniques, except where some explanation is essential to an understanding of the text. A recent book, *Methods in Plant Ecology* (Chapman 1976), covers this field. My book is basically concerned with the individual plant, and is consequently autecological and does not attempt to explore the many facets of community and production ecology.

In conclusion, I hope that this book provides a novel synthesis of physical and ecological information.

Acknowledgments

Acknowledgments and thanks are due to the many persons and organizations who have permitted material to be used in this book. Firstly, I should like to thank the numerous individual authors of the various publications who have provided the raw material for this book. Secondly, officers of the following societies and organizations are thanked for permission to reproduce specific items from their publications. These include: *Berichte der deutschen botanischen Gesellschaft* (Fig. 6.16), Edward Arnold (Fig. 2.6), *Flora, Jena* (Figs. 3.12, 4.6, 4.7, 4.8, 4.20), Harper & Row (Figs. 2.1, 2.4, 2.5), Harvard University Press (Fig. 2.3), *Journal of Ecology* (Figs. 2.9, 3.10, 4.2, 4.15, 5.8, 6.8–9, 6.13, 6.19, 6.21, 7.1, 7.5–8, 7.12–13, 7.16, 8.2), *New Phytologist* (Figs. 7.9, 8.4), *Physiologia Plantarum* (Fig. 6.15), *Plant & Soil* (Fig. 7.14), *Science Progress* (Figs. 3.4, 3.5) and Verlag Eugen Ulmer (Figs. 3.6, 4.10, 5.2, 5.7).

Finally, I should like to give special thanks to those colleagues, advisers, secretarial staff and, last but not least, my long-suffering wife and family who have helped and encouraged me during the preparation and completion of this book.

1: Introduction

I.I WHAT IS PHYSIOLOGICAL PLANT ECOLOGY?

Physiological plant ecology is a combination of plant physiology with plant ecology and therefore is concerned with the nature, or functioning, of plants in relation to their environment.

Unfortunately physiology and ecology have tended to deviate from each other. In order to understand the functioning of plants, physiologists have investigated the separate effects of individual environmental parameters and have become increasingly concerned with changes that take place at the cellular and molecular levels of organization. This can lead to divorcement from environmental realities although it is perfectly possible to conduct biochemical studies which have an ecological relevance (e.g. Crawford 1966, Woolhouse 1969). On the other hand, the ecologist may concentrate his attention on either the plant or the environment at the expense of the interaction. The phytosociologist is concerned with groupings of plants and the elegance of taxonomic (e.g. Braun Blanquet 1932) or mathematical techniques (e.g. Kershaw 1973) may divert attention from environmental relationships while too great an emphasis on the environment may lead to the treatment of plants merely as troublesome complications in the formulae of environmental chemistry and physics.

Physiological plant ecology, autecological as it is, is concerned with the response of the individual plant to its environment and the ways in which changes of environment are accommodated by the reactions of the plant. There is an erroneous tendency to think of physiological-ecological studies as a modern synthesis but they are as old as the subject of plant ecology itself. The first use of the term 'oecology' is attributed to the zoologist Haeckel (1869) and was subsequently defined by him as the study of the reciprocal relationships between organisms and their environment. The first general textbooks on plant ecology were those by the Danish botanist

1

Warming (1896) and the German Schimper (1898). The latter book
was the first to appear in English translation (1902) and its author
was well aware of the importance of physiological studies for he
states (in translation)

'. . . The ecology of plant distribution will succeed in opening out
new paths only on condition that it leans closely on experimental
physiology, . . .'

He proved his point by devoting at least 200 pages within his book
to environmental factors and plant reactions. This book addresses
approximately the same number of pages to similar topics.

1.2   PLANT RESPONSES TO THE ENVIRONMENT

Plants respond both to the physical environment and to each other.
The mutual interactions may be influenced by the form and physiology of the particular plant species, be modified by the environment
and may in turn alter the environmental response. Plants may
also vary in their responses on a seasonal basis and show different
reactions at various stages in their life history.

For example, within woodland, spatial differences provide habitats
beneath, between and upon the trees; local variations in the availability of light, water and nutrients cause further differentiation that
can be exploited by plants with different functional abilities; on a
seasonal basis, some plants (e.g. bluebell, *Endymion non-scriptus*) are
adapted to grow and develop while the trees are still leafless, whereas
others (e.g. ferns) are shade-tolerant and develop later.

Considerations such as these are involved in the concept of *niche*.
A plant's niche is an abstraction compounded of its spatial, functional
and temporal relationships within a community. The niche of one
species may overlap with those of others, but the evolution of
differences in niche minimizes interspecific competition and allows
the development of a complex community such as woodland.

1.2.1   *Responses to environmental factors*

All plants have certain basic requirements without which they cannot
exist. Each factor usually has a minimum and a maximum level
beyond which a plant cannot survive. However, the extreme values
tolerated by one plant are not necessarily identical to those tolerated

by another. Consequently the study of the tolerance of plants to environmental extremes may provide a means of discriminating between plants with different ecologies. The optimum response is somewhere between the two extremes and may also differentiate between species. There is commonly a dissimilarity between laboratory and field response (Chapter 8), as the latter is modified by competition. The overall relationship of a species to its natural environment is its ecological amplitude. Species usually have separate but overlapping amplitudes (Figure 1.1), those that appear to have identical amplitudes may make different demands upon the environment as in the case of a forest tree and a woodland herb. Species with wide amplitudes may have a great phenotypic plasticity but more commonly are composed of ecological races (ecotypes) with a range of overlapping amplitudes (Figure 1.2).

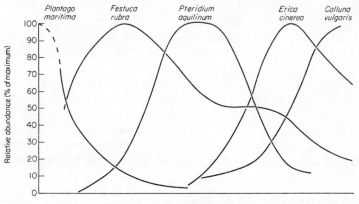

FIGURE 1.1. Overlapping ecological amplitudes in species from cliff-top communities (based on unpublished data from the Isle of Man).

The species are ordered with respect to the vegetational gradient by the continuum analysis of Curtis & McIntosh (1951).

Environmental factors can be correlated with the vegetational gradient so that the peak of *Plantago maritima* is found to be associated with proximity to the sea, and organic soils of high pH, conductivity (i.e. salt content) and base status: *Calluna vulgaris* is found at a greater distance from the sea on organic soils of low pH and base status with an intermediate conductivity; whilst the peak of *Pteridium* is at intermediate altitude on less organic soils of low conductivity and cation exchange capacity with an intermediate pH and base status.

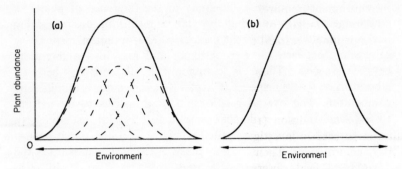

FIGURE I.2. Identical ecological amplitudes within a species caused by (a) Three overlapping ecotypes and (b) one ecotype with great phenotypic plasticity.

I.2.2   *Modification by intrinsic (plant) factors*

Plants modify their environment. Vegetation ameliorates the extreme conditions that would otherwise exist at the soil surface, producing a microclimate which is shaded, humid, temperate and free of wind and consequently suitable for the growth of other plants (e.g. mosses) that are adapted to exist in such conditions. Plants may also modify their environment in a way which furthers their own growth and excludes other species as in the acidification of chalk grassland by bushes of heather and gorse (Grubb *et al.* 1969). Thus, species that are adapted to calcareous soils are excluded from the acidified areas, similarly species that are not adapted to shade are unable to exist beneath other plants. Exclusion may be more direct. In acid pastures the heath rush (*Juncus squarrosus*) and the mat grass (*Nardus stricta*) have squarrose habits which enable them to suppress adjacent plants. Such aggressive competitors are able to spread at the expense of their rivals even when both are adapted to a similar environment.

The stage of growth may also modify the response. Species with comparable ranges of tolerance may differ in their seasonal reactions —for example the rate of hardening in autumn or increase of susceptibility in spring may differentiate between two species of comparable frost resistance. Young plants, especially seedlings, may be more sensitive or show a different response to environmental factors than the mature plant and thus the distribution of mature plants may be determined by the tolerances of seedlings.

Consequently, the examination of physiological responses to the environment requires an appreciation of the environment, the tolerance ranges of species and the modifications exerted by competition, seasonal changes and stage of growth.

1.3   APPROACHES TO THE SUBJECT

Grime & Hodgson (1969) have considered various strategies that can be used in investigating the determination of plant distribution by environment. Their three basic methods are: a direct approach based on field observations, comparisons of environments and comparison of plants.

1.3.1   *The direct approach*

This involves the careful examination of the reasons for the failure of a species to transgress a vegetation boundary (e.g. Grime 1963) and is only valid when environmental factors, rather than chance distributional factors, are known to determine the distribution of vegetation. Frequently the boundaries between areas of vegetation are artificial and depend upon past management. At the best the method is slow and laborious: seedlings may persist for several seasons before they expire and may only cross the boundary at all in an exceptional season. The process can be accelerated by the artificial introduction of propagules (cf. Putwain & Harper 1970) or by deliberately modifying the environment (Harper 1967).

1.3.2   *Comparison of environments*

The environment in which a particular vegetation type or plant occurs is compared with those in which it is absent; or the abundance of a species is related to the magnitude of various environmental factors. The technique requires the measurement of a large number of environmental variables in order to find those which show differences. Unfortunately environmental variables are often intercorrelated and although techniques such as multiple regression, multiple correlation (Table 1.1) and principal components analysis (Gittins 1969) may simplify or summarize the relationships they may still not permit a causal interpretation. An association of differences in vegetation with changes in a particular environmental variable may suggest

TABLE I.I. Intercorrelation of environmental factors.
The following table is derived from selective multiple regression analyses
of the performance of heather (*Calluna vulgaris*) in S.W. Scotland (Thorp
1972).

---

1 *Primary regression.* *Calluna* performance was related to

   altitude (−)   longitude (−)

---

2 *Secondary relationships.* The following soil variables are related to

   (a) altitude   (b) longitude
     % ignition loss (+ −)   % ignition loss (+)
     peat depth (+ −)   exchangeable K (− +)
     exchangeable K (−)   exchangeable Na (− +)
     exchangeable Na (− +)   exchangeable Ca (− +)

---

3 *Further relationships.* The above variables are interrelated and also related to
further soil and environmental variables.

Variable	Related to
peat depth	slope (−)
ignition loss	slope (−), TEB (−), exch. K (+) exch .Ca (+)
exch. K	exch. Na (+), ignition loss (+ −)
exch. Na	exch. K (+), ignition loss (+)
exch. Ca	exch. Na (+), ignition loss (+)

---

4 *Conclusions.* The relationship of *Calluna* performance to altitude and longitude
is associated with variation in at least seven other variables. The measurement
of climatic and microclimatic factors would have introduced many further
correlations.

Notes Relationships have been indicated as follows: (+), positive correlation,
(−) negative correlation, (+ −) curvilinear relationship with an initial rise
followed by a fall (− +) curvilinear relationship with fall followed by a rise.
TEB= total exchangeable bases.

causality although other correlated variables may be responsible.
Equally, the vegetation may have caused changes in the environment,
or historical factors such as previous management and the conditions
prevailing during the early life history of the plant may have obscured
the causal relationship between the environmental factors and the
vegetation. At the best, comparison between environments is a
guide to possible causation which can then only be determined by
experiment.

1.3.3 *Comparison of plants*

One of the difficulties of comparing plants is the selection of the parameters of performance that are to be measured. Growth is often used, but this may not be appropriate in interspecific comparisons where patterns of growth may be inherently dissimilar (Rorison 1969). Grime & Hodgson (1969) have recommended the use of large-scale surveys in order to select features common to a large number of species. Alternatively, it is possible to compare ecotypes (e.g. Hutchinson 1968, 1970a,b; Proctor 1971a,b) where the number of inherent differences is less.

Comparisons usually consist of growing each type in its own environment and in that of the other species. A complete simulation of environments may give no more information about causal factors than field studies and the experimenter usually attempts to regulate certain environmental factors in order to assess their influence. It may prove difficult to select the appropriate factors from the environmental complex. For example, plants subject to iron chlorosis may fail to establish themselves on calcareous soils because of alterations in their water relations rather than directly on account of iron deficiency (Hutchinson 1970a,b).

1.3.4 *Combined approaches*

It is obvious that there is no unique approach and investigators usually adopt a combined method when they approach a specific problem in physiological plant ecology. The initial level of investigation depends upon the amount of information already available. If there is no prior knowledge then a possible sequence would be to initially correlate vegetation and habitat in the field, then use any correlations as a basis for laboratory comparisons designed to determine the causal factors, and finally to test laboratory findings by re-examining their impact in the field. A possible strategy is outlined in Figure 1.3.

The consequence of such a strategy, and also of other approaches, makes it essential for the physiological plant ecologist to have a knowledge of the workings of environmental factors as well as an appreciation of the physiological responses of plants. The following chapters are written with this in mind.

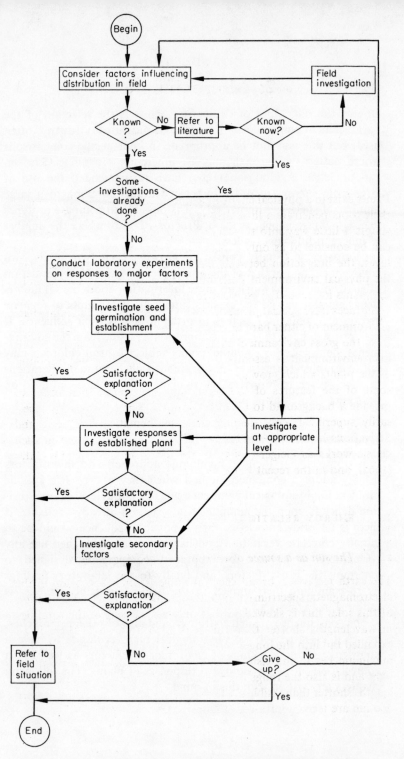

FIGURE 1.3. A simple strategy for investigation of the physiological plant ecology of a species or group of species. (Bannister 1976).

2: Microclimate

Plants exist in a physical environment near the surface of the ground; their roots penetrate a little distance into the earth's crust and their shoots a little way into the atmosphere, but on a global scale they can be considered as only superficial. However, within this narrow layer, the interaction between plant and environment is complex: the physical environment determines the plants which grow, whilst the plants influence the physical environment.

Surfaces have special properties with the result that the immediate environment of either bare or vegetated surfaces is markedly different from the gross environment that surrounds them. The study of this microenvironment is essential to an appreciation of the responses of the plants which grow in it. This chapter sets out to summarize some of the features of the aerial microenvironment in order to provide a background to Chapters 3 and 4. The treatment is necessarily superficial, consequently most mathematical formulae and derivations have been omitted. A fuller treatment is found in such classic works as Geiger (1965), in the useful monograph by Gates (1962), and in the recent book by Monteith (1973).

2.1 ENERGY RELATIONS OF SURFACES

2.1.1 *The sun as a source of energy: the radiation balance*

The earth receives a broad spectral range (from 0·2–100 μm in the electromagnetic spectrum) of radiation from the sun. The distribution of this solar flux is skewed so that only 25 per cent of the total lies in wavelengths shorter than the peak (470 nm) whilst there is an extended tail into the longer wavelengths (Figure 2.1). The radiation maximum lies in the visible range (360–760 nm) which constitutes light and is also the range over which photosynthesis occurs. Wavelengths shorter than visible are termed ultra-violet whilst those longer 760 nm are termed infra-red (Figure 2.1).

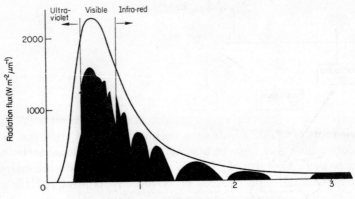

FIGURE 2.1. The solar radiation flux as approximated by the theoretical curve for a black body radiating at 6 000°K (upper line) and the flux received at the earth's surface (shaded area). Based on Gates (1962). Absorption is mainly due to ozone in the visible and ultra-violet, by oxygen in the visible, and water vapour in the visible and infra-red. Water vapour ozone and carbon dioxide also absorb radiation in the further infra-red (Figure 2.4).

The integrated flux density of solar radiation incident upon a surface perpendicular to the sun's rays at the outer edge of the earth's atmosphere is about 1,400 W m^{-2} (2·0 cal cm^{-2} min^{-1}); this is the solar constant. However, in an average year in the northern hemisphere, less than half this radiation reaches the earth's surface; the rest is reflected, scattered and absorbed in the atmosphere (Figure 2.2). The amount scattered and absorbed is dependent upon solar angle, upon absorption and scattering coefficients which are complex functions of wavelength and upon the depth, density, composition and temperature of the atmosphere. Theoretical relationships are given by Gates (1962) but estimates of the amounts scattered and absorbed are more readily obtained from empirical equations derived from direct observations (e.g. Blunt 1932). The absorption of longer wavelengths is largely by carbon dioxide and water vapour whilst oxygen, ozone and nitrogen absorb much of the ultra-violet. Some of this absorbed radiation is re-emitted at longer wavelengths and is termed 'counterradiation' (c) and is sufficiently distinct from the shorter wavelengths of solar radiation to be distinguished as 'longwave' as opposed to 'shortwave' radiation. The counterradiation

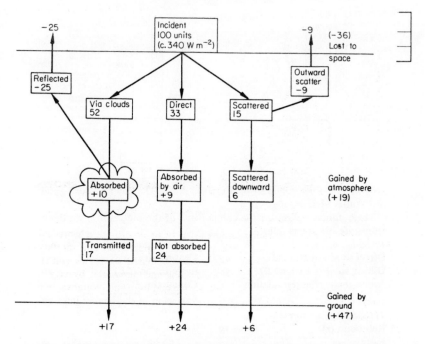

FIGURE 2.2. The passage of solar radiation through the atmosphere. (Average annual values for the Northern Hemisphere—after Gates 1962).

is an important contributor to the total thermal energy received at the earth's surface (Figure 2.3) and is the main source of energy at night. The shortwave input during the day consists of both direct solar radiation (I) and nondirectional sky radiation that has been scattered and diffused by the constituents of the atmosphere. In the visible spectrum these components are recognized as direct sunlight and diffused daylight. The total ($I + D$) is termed global radiation.

At the surface some of the radiation is lost by reflection (r) and yet more (b) is reradiated at a longer wavelength in accord with the Stefan-Boltzmann Law which relates the total intensity of radiation emitted from a 'black body' (i.e. a body which completely absorbs all radiation incident upon it) to the fourth power of the absolute temperature (T). Thus

$$b = \sigma T^4$$

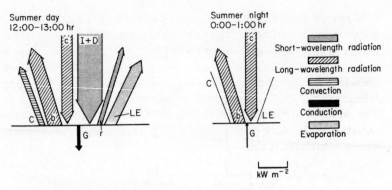

FIGURE 2.3. Energy budget of a surface during the day and during the night (after Geiger 1965).

Mean annual values for the components of the budget in a northern temperate site are as follows.

Radiation balance	W m^{-2}	Other componenets	W m^{-2}
Direct solar radiation (I)	45	Heating air (C)	-6
Diffuse solar radiation (D)	58	Evaporation (LE)	-40
Atmospheric counterradiation		Conduction (G)	0*
(c)	319	Balance	-46
Terrestrial radiation (b)	-357		
Reflection (r)	-19		
Balance (S)	$+46$	*net (gains and losses cancel out)	
Other components	-46		
Overall balance	0		
($S + LE + C + G = 0$)			

where σ is a constant of proportionality (5.67×10^{-8} Wm^{-2} $^{0}K^{-4}$). Any terrestrial objects are reasonable approximations to black bodies and thus outgoing radiation may be estimated by the Stefan-Boltzmann equation. The longwave radiation flux (R) is the difference between the amount gained (conventionally given a $+$ sign) by the surface from the atmospheric counterradiation (c) and that lost by black body radiation (b) (conventionally negative). Thus

$$R = c - b$$

and the total radiation balance (S) may be given as

$$S = I + D + R - r.$$

The various quantities are given positive signs when they represent net gains to the surface and are negative for net losses.

Much of the outgoing longwave radiation is trapped by the atmosphere, contributing to the so-called *greenhouse effect* which results in the warming of the air. However, in the region between 8 and 13 μm the absorbtion of radiation by water vapour and carbon dioxide is ineffective and just beyond 4 μm it is zero, thus providing 'windows' by which the radiation can escape to space (Figure 2.4). The window around 10 μm is the most important for longwave radiation as the black body radiation for the earth and its atmosphere have a maximum at this wavelength.

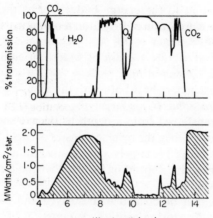

FIGURE 2.4. The transmission and emission of longwave radiation by the atmosphere (after Gates 1962).

Above: Transmission—note the high transmission at 4 and 5 μm and between 8 and 13 μm, these constitute 'windows'.

Below: The observed downward emission by the atmosphere at night. The gap in the region between 8 and 13 μm is due to the 'window' allowing losses from the atmosphere to space. A similar, but smaller, loss occurs around 4 and 5 μm, but the atmospheric emission is relatively low at these wavelengths.

2.1.2 *The energy budget (Figure 2.3)*

The basic components of the energy budget remain the same for the surface of an object as large as the earth or as small as a leaf. There are four major components which are best considered as fluxes as they are sometimes positive and input energy to the system whilst at other times they are negative and represent an energy loss. The *radiation flux* (or radiation balance), *S*, has a positive value during the day when solar energy is available directly but is negative at night when the black body radiation from the surface exceeds the counterradiation.

$$\text{i.e. } (-)\, S = c - b,$$

The radiation flux represents the principal input of energy: the *evaporative flux*, *LE* (where *L* is the latent heat of evaporation and *E* the evaporation rate), often represents the main loss. This relationship of evaporation to the energy budget forms the basis of most formulae for the calculation of irrigation requirements from meteorological data (e.g. Penman 1948). At 15°C, almost 2 500 J (589 cal) are required to evaporate 1 gram of water, an equivalent amount being gained on condensation.

The surface exchanges heat both with the ground and with the air. The *ground-surface flux* (*G*) is largely by conduction and is positive when heat is transferred from a depth to the surface (as on a cold night) and negative when the reverse transfer occurs (during the day). The *surface-air flux* (*C*) is largely a convective transfer and may also be positive or negative depending on whether the air passes heat to the surface or vice versa.

As energy cannot be created or destroyed and the surface itself has no capacity for storing heat, the summation of all the fluxes is always zero:

$$S + LE + G + C = 0$$

or, expanding the radiation flux, *S*

$$I + D + R - r + LE + G + C = 0.$$

The basic equation can be modified. An important omission, but a very difficult component to measure is the lateral transfer of energy by gross movement of the air (i.e. wind or *advection*) whereby warm or cold air may flow past the surface and add or abstract heat. Precipitation may provide an input or loss of heat depending upon its temperature. When solid objects rather than surfaces are under consideration, then a storage factor for heat is also included.

Figure 2.3 gives a diagrammatic representation of the energy budget. During the day the main input is from shortwave radiation and the main loss through evaporation; at night the longwave radiations are of greatest importance with the main gain through counterradiation and the main loss through radiation from the surface. Thermal energy is lost by the surface to both air and sub-surface layers during the day $(-G, -C)$, but the situation is reversed at night $(+G, +C)$.

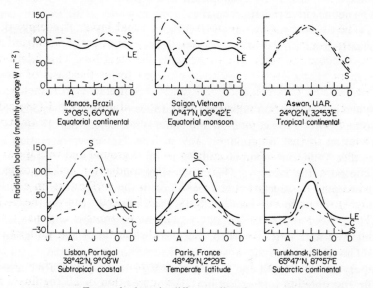

FIGURE 2.5. Energy budgets in different climatic zones (Gates 1962).

The relative contribution of the various components of the energy budget varies both with season and geographical position (Figure 2.5). In close proximity to the equator there is no great annual fluctuations in the radiation balance except with respect to variations in cloudiness, thus in Saigon in April a maximum is obtained before the commencement of the monsoon. At higher latitudes there is a distinct summer maximum with a winter minimum which may even be negative (e.g. Paris, Turukhansk). The ground-surface flux (G) is the smallest of the other components and has hence been omitted in Figure 2.5, although in certain circumstances it may represent the main source or sink of heat. Thus in winter at high latitudes, most of the incident radiation may be used to heat the ground, whilst on a short summer night the energy gained by the ground during the day may be the main source of heat. The remaining components, the evaporative flux (LE) and the surface-air flux (C), show an inverse correlation. In seasons when water is relatively unavailable (i.e. throughout the year at Aswan in the desert, before the monsoon in Saigon, in the European summer, and whilst all water is still frozen in the Siberian spring), the heat losses are mainly due to atmospheric convection, otherwise evaporation is the main source of heat loss. The net negative radiation balance found in winter at higher latitudes is compensated for by the abstraction of heat from the atmosphere, usually from winds that are slightly less cold than the ground.

2.1.3   Vegetation and radiation

When solar radiation falls upon the surface of a stand of vegetation, some is lost to the vegetation and some is reflected. This is a similar situation to that obtaining at any surface. However, vegetation is neither solid nor opaque and some of the shortwave vegetation penetrates the canopy. The penetrating radiation is increasingly intercepted by vegetation at successively lower levels but some may pass through to the ground. Both the radiation that reaches the ground and the radiation intercepted by vegetation will be part reflected and part absorbed, but leaves also transmit radiation.

The relative amounts absorbed, reflected and transmitted depends upon the wavelength and there is a strong similarity between curves for transmission and reflection (Figure 2.6). Maximum values are found in the near infra-red (750–1 500 nm) where the absorption is low and there are also secondary, weaker, peaks of reflection and transmission in the visible spectrum between 500 and 600 nm in the yellow-green range. Cabbage plants investigated by Moss & Loomis (1952) also show an appreciable reflection of violet light (400 nm). The peak of transmission in the yellow-green range accounts for the greenish colour of light under vegetation, although if the eye were sensitive to the near-infra-red this 'colour' would swamp all others. The reflection of infra-red is apparent in photographs made with film sensitized to the appropriate wavelengths where vegetation appears light-coloured. However, most of the visible and far infra-red light is absorbed; the former range is used in photosynthesis although the amount is negligible (< 5 per cent) and both contribute to the heating of the leaf and the evaporation of water. Heat is also lost by longwave radiation from the ground and vegetation and may be gained from counterradiation.

There is a general attenuation of unabsorbed radiation as the ground surface of a stand of vegetation is approached. The reflection and reradiation of energy from the depths of the stand will be weak, not only because of the small input at these lower levels but also because much of the escaping radiation is intercepted by the upper layers of the vegetation. It follows that the largest exchanges of energy will occur at or near the surface of the vegetation and that many of the correlated microclimatic factors will show their greatest fluctuations at the same levels.

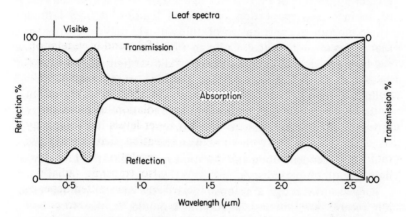

FIGURE 2.6. Idealized relationships between reflection, transmission and absorbtion with respect to wavelength for a green leaf (after Monteith 1973).

2.2 MICROCLIMATE IN RELATION TO ENERGY BUDGET AND VEGETATIVE COVER

The component fluxes of the energy budget of a surface can show large alterations in both direction and amount. These fluctuations are manifest as changes in readily observable environmental factors such as light, temperature, moisture and air movement. All of these factors have profound effects upon the responses of plants and, moreover, are themselves modified by the presence of vegetation. The responses of plants to environmental factors are dealt with in subsequent chapters: the rest of this chapter deals with the factors themselves and the way in which they are modified by the presence of vegetation.

2.2.1 *Light as an environmental factor*

Light is the portion of the shortwave radiation spectrum that is encompassed by the range between 360 and 760 nm and as such may be treated in a similar manner to the solar radiation balance (2.1.1). Thus the light that is incident upon a surface is either direct sunlight or diffused light that has been variously scattered by particles and droplets in the atmosphere. Light is also absorbed by the atmosphere so that at high altitudes there is a more intense illumination with a greater proportion of short wavelengths. Low solar angles contribute to an increased proportion of diffused lighting at high latitudes where the annual contribution of diffused light may be greater than that of direct sunlight (e.g. Collman 1958). A low solar angle below the horizon leads to the persistence of diffuse light after sunset (or its occurrence before sunrise) and the long transitional period of twilight in high latitudes can be contrasted with the abrupt transition in the tropics where the sun attains a high elevation at noon and sets more or less perpendicular to the horizon. Twilight may be of little relevance to the photosynthesis of plants but is important in responses that are dependent upon the duration rather than the intensity of light (e.g. photoperiodism).

Angled surfaces that are sloped towards the equator will theoretically receive more direct insolation and should be, therefore, better illuminated, warmer and drier than poleward slopes. Theoretical relationships are given by Gates (1962) whilst Kaempfert (1942) and Kaempfert & Morgen (1952) present tables and graphs dealing with a variety of angles of slope, aspects and seasons. However, the contribution of diffused light may be considerable so that, while the amount of direct insolation on a north-facing slope of 30° in southern Bavaria is only 2 per cent in winter and 73 per cent in summer of that of a comparable south-facing slope, the actual values for global radiation were 32 per cent and 94 per cent respectively (Grunow 1952). A further complication is provided by obstructions (Figure 2.7), thus mutual shading by mountains can mean that sunward slopes actually receive less radiation than those with a slight poleward inclination (Green 1964). The increased contribution of diffused light in cloudy oceanic climates tends to negate the effects of aspect upon the distribution of vegetation although detailed analyses may still reveal differences. Thus, on isolated mountains in north-western Scotland, species of northern affinity such as *Vaccinium*

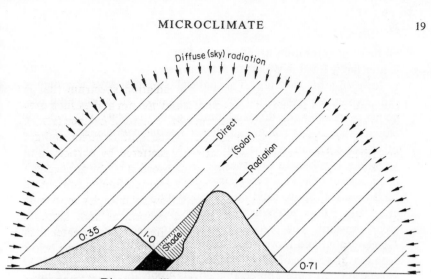

FIGURE 2.7. Diagram to illustrate the contribution of direct and diffuse radiation to a terrain of varied topography.
Figures indicate relative intensities of direct radiation on various slopes.

vitis-idaea and *Empetrum hermaphroditum* are more abundant on north-facing slopes whereas *Erica cinerea*, of more southern affinity, is more abundant and earlier to flower on south-facing slopes (Gimingham & Cormack 1964). In heathland in south-western Scotland, *E. cinerea* shows the same propensity and the proportion of bryophytes increases on slopes with a more northerly aspect (Thorp 1972). However, it should be remembered that correlations of vegetation with aspect are not wholly due to changes in the radiation climate as the direction of the prevailing wind will influence both evaporation rates and the distribution of rainfall. Accordingly, the prevailing south-westerly winds may also contribute to these differences between north and south-facing slopes in Scotland. In climates where moisture or radiant energy is in short supply, aspect becomes important: hence in polar regions the scant vegetation is found mainly on slopes with an equatorial aspect whilst in arid regions nearer the equator only the poleward slopes may retain enough moisture to permit some plant growth.

The *albedo* (reflectivity) of natural surfaces may be of importance in determining the light and energy regime of plants growing on or near these surfaces. Dark soils have low albedos (< 10 per cent) and all types of vegetation usually have albedos of less than 30 per cent. Water has a low albedo (3–10 per cent) and the albedo of a surface is lowered after wetting. The natural surfaces with the highest reflectivities are fresh snow cover (c. 95 per cent), dense cloud (60–90 per cent) and light, dry sand dunes (30–60 per cent). The radiation climate above snow and in sand dunes is extreme with a large amount of reflected radiation. The plant, or the human, in either of these situations is subjected to insolation both from above and below and the resultant heat load can be excessive. Gates (1962) calculates the load on a leaf above sand on a clear summer's day to be c. 2·4 cal cm^{-2} min^{-1} (c. 1 700 Wm^{-2}), a value 20 per cent greater than the solar constant. The preceding values for albedos are taken from Geiger (1965) and more are available in Penndorf (1956). It can be seen that the range of albedos is such that a difference in land use or vegetation can have appreciable effects upon the radiation climate of an area.

2.2.2   The modification of the light climate by vegetation

The visible spectrum coincides with wavelengths that are most active in photosynthesis and thus the fate of light within a canopy of vegetation is of interest to the ecologist. The reduction of light intensity leads to a lowered potential for photosynthesis and alterations in spectral composition may also have some bearing on the responses of plants growing in shade.

The pattern of attenuation of light is dependent upon the growth form of the plants which make up the canopy. Plants with a dendroid habit (e.g. trees, shrubs, dwarf shrubs, many herbs and certain mosses) will reduce light intensity rapidly within their crown regions whilst swards of grassy species will show a more gradual reduction of light intensity with depth. Baumgartner (1955) has investigated the relationship between the attenuation of light and growth by plotting relative height of the vegetation against the relative light intensity found at any particular level. A similar approach is used in Figure 2.8 which is derived from data of Newton & Blackman (1970). The theoretical basis for the attenuation of light within a canopy has been provided by Monsi & Saeki (1953) who present a relationship analogous to the Beer-Lambertson Law:

$$\text{i.e. } I = I_0 e^{-KL}$$

where I_0 is the intensity of the incident radiation, I is the intensity after passage through a canopy of a given leaf area index $(L)^*$, whilst K is the *extinction coefficient*. The mathematical model assumes that a constant proportion of light is intercepted by a given leaf area. This condition is not fulfilled in practice. The orientation

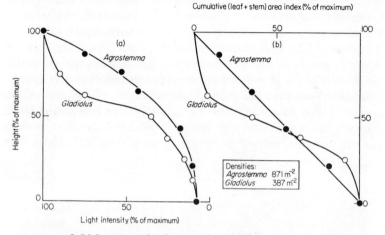

FIGURE 2.8. Light attenuation by canopies of different structure. (Derived from Newton & Blackman (1970), cf. Baumgartner 1955).

(a) Decline in light intensity. Note the exponential decline in *Agrostemma* which has horizontally oriented leaves and the initially poorer interception by *Gladiolus* with its vertically oriented leaves.

(b) Change in cumulative leaf + stem area index. Note the similarity in pattern to (a).

of leaves and light source, the spectral composition of the light and its various absorbtion, transmission and reflection by leaves of different morphology and anatomy (e.g. sun and shade leaves) all contribute to variations in the value of K. Thus, Newton & Blackman (1970) show that *Agrostemma githago* (corncockle), an almost ideal plant from a theoretical point of view as it shows little change of stem and leaf area index with height and has horizontal leaves, has an extinction coefficient of *c*. 0·4 whilst a dense population of *Gladiolus*, with linear and erect leaves, gave an extinction coefficient of only 0·15 (see also Figure 2.8). However, populations

*Total area of leaves above a unit area of ground.

with low values for K may still show a considerable attenuation of light if their combined leaf and stem area indices are high (Anderson 1966a,b).

The main changes that occur in spectral composition are related to the restriction of the shorter wavelengths. Although there is a general attenuation of shortwave radiation, the green and particularly the near infra-red are the least reduced by vegetation. Consequently, the ratio of visible to total shortwave radiation decreases with increasing leaf area index (Figure 2.9), the steepest decline being found in plants such as kale with horizontally orientated leaves. The interception by a moderately dense stand of *Gladiolus* (Newton & Blackman 1970) results in almost complete extinction of visible light while there is only an 80 per cent attenuation of the near infra-red. Thus the near infra-red accounts for 78 per cent of the total radiation at ground level as against 51 per cent in the open. As the ratio of visible to near infra-red radiation decreases, so does the ratio of red (660 nm) to far red (730 nm) (R/FR). This shift in

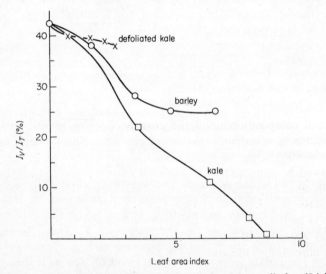

FIGURE 2.9. The ratio of visible (I_v) to total shortwave radiation (I_T) in canopies of defoliated kale, barley and kale in relation to area indices (Szeicz 1966).

The greatest changes in spectral composition are shown within stands of kale where leaves are oriented horizontally (c). A decrease of I_v/I_T will indicate a corresponding decrease in the red/far red ratio.

R/FR ratio should have important repercussions on the germination, establishment and development of plants growing in the shade of others (4.1). There appear to have been few ecological investigations conducted from this standpoint.

The vegetation canopy is penetrated by both direct and diffuse light. The penetration of both is related to the structure of the canopy, but the direct component is the most difficult to measure. More light may penetrate a stand of vegetation when the solar angle is low and the light passes laterally through the trunks than when the sun is at a high elevation and direct light has to pass directly through the canopy. However, the point to point reception of direct sunlight is very variable as the spatial distribution of the canopy may be disturbed by wind and the sun obscured by cloud. The contribution of direct sunlight through small holes in the canopy (sunflecks) may be quite considerable and form the major part of the incident shortwave radiation (Evans 1956; Whitmore & Wong 1959). Long-term averages smooth out some of the variation in direct lighting but over the long term the vegetation may also show periodic changes as in a deciduous forest. Anderson (1964, 1966b) has used hemispherical (180°) photographs taken with a camera with a fish-eye lens to interpret both the direct and indirect components at various times of year.

The ecological significance of some of the variations in light climate is insufficiently investigated. Little is known about either the contribution of sunflecks or the effects of light of different spectral composition upon the growth of shade plants. *Impatiens parviflora*, a woodland annual, is similar to most plants in that photosynthesis, on an energy basis, is most efficient in red light, followed by blue and then green (Hughes 1965)—despite the preponderance of greenish light beneath vegetation. However, excessive amounts of either red or blue light cause morphological changes such as undue hypocotyl extension (red) and restricted leaf expansion (blue) which could be disadvantageous in a shaded habitat. Other adaptations to shade are better known: plants may avoid the effects of shade by rapid growth when light is available or by growing past the shading vegetation and they may persist until the canopy opens up, aided by modifications such as seeds with a large food reserve, low growth and respiration rates, low compensation points and resistance to fungal attack (Grime 1966). Such physiological adaptations are discussed more fully in Chapter 4.

2.2.3 *Temperature profiles near a bare surface*

Temperature is the qualitative measure of the molecular activity caused by a quantitative change in the heat content of a body. (The calorie is the most well-known unit of heat, but the corresponding SI unit is the joule (J) with 1 cal = 4·185 J).

The large changes in energy balance that occur in the vicinity of a surface are reflected in large temperature fluctuations. The surface itself shows the maximum gains and losses in energy and consequently the largest fluctuations in temperature. In the lowermost atmospheric layer (the troposphere), the atmosphere well above the ground shows a general decrease of temperature with height; the amount of heat retained by the atmosphere becoming smaller as the atmosphere becomes less dense. Converse relationships may be found in other atmospheric layers but these are of no direct importance to plants. Figure 2.10 shows the temperature gradients above and below a soil

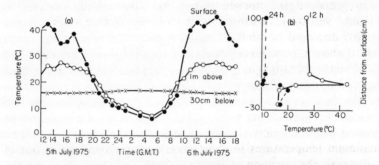

FIGURE 2.10. Diurnal course of temperature at, above and below the ground surface. (a) Course of temperature at three levels. (b) Temperature profiles at midnight and noon. Dunblane: Perthshire. (The assistance of Paul and Katy Bannister is gratefully acknowledged.)

surface during still and clear weather. During the day there is typically a marked fall-off of temperature with height above the ground and with depth below. During the night the minimum temperature is found at the surface of the ground and there is an increase of temperature with height until the normal lapse rate is re-established. This is known as a temperature inversion. The maximum temperature at night is below the ground due to slow conduction from the surface although at depth a more or less constant soil temperature is attained. Thus the plant grows in an environment that is much more extreme than that indicated by normal measurements of air temperature. These are usually made in a Stevenson screen with a thermometer 4 ft (1·2 m) above the ground although British Meteorological Stations often measure the 'grass minimum' temperature and make the distinction between 'ground' and 'air' frosts. The more extreme temperatures that develop near the ground make ground frosts more common than air frosts and of greater interest to the gardener, farmer and plant ecologist.

2.2.4 Temperature profiles within vegetation

The distribution of temperature in the vegetation canopy depends not only upon the heat budget but also upon the structure and disturbance of the canopy. The radiation climate of grassland with vertically orientated leaves differs from that of a stand of broad-leaved herbs which is again different in some ways from that of a forest.

In grassland the upright orientation allows bodies of air to migrate vertically; mixing is aided by the fact that grasses are readily displaced by wind. Thus in grassland the effective surface is a little above ground level. Measurements made during the afternoon by Waterhouse (1955) in a meadow with grass 0·5 m tall, indicate a marked decrease of temperature below 30 cm where the maximum occurs. In short grass the maximum is near ground level, as it would be with bare ground, but as grass grows the maximum is found above ground but well below the upper surface of the canopy. At night, minimum temperatures in short grass are at ground level, but in taller grass the retention of a warmer layer near the ground and the downward migration of cold air from the surface of the vegetation result in a minimum being found at some height within the stand. These points are very well illustrated by the early work of Geiger in stands of rye (see Geiger 1965). Temperatures within tall grass are never as extreme as those developed over bare ground or short grass (e.g. Norman et al. 1957).

In contrast, the maximum heat exchange of herbaceous vegetation with horizontally oriented leaves takes place at the upper surface of the vegetation. Consequently maximum temperatures are found near the upper surface of the canopy during the day. The nightly minimum is found at ground level because the vegetation does not present a barrier to the downward migration of cold air under the influence of gravity and mixing is prevented by the structure of the canopy.

In a dense stand of trees (e.g. Baumgartner 1956) the vegetation canopy provides an effective barrier between ground and air. Maximum temperatures are found in the upper crown area in the early afternoon and there may then be a difference of several degrees between the canopy and the ground (Figure 2.11). The maximum temperature at ground level is likely to be achieved later in the afternoon. At night the most negative radiation balance occurs in the crown region, although some exchange with the ground does occur, so that there is little variation in the stratification of temperature. In a sparse stand of trees (e.g. Göhre and Lützke 1956) the vertical distribution of temperature may be less different from that of the open ground and there may be marked temperature inversion at night with little vertical stratification during the afternoon as the natural increase in air movement is relatively unimpeded by the more scattered trees.

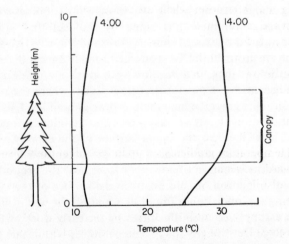

FIGURE 2.11. Temperature profiles in a stand of young spruce at the warmest and coldest parts of the day and night. (Derived from Baumgartner 1956.)
Note the heat retaining capacity of the canopy, particularly during the day.

Sun of a low solar angle may penetrate a stand laterally and the south sides of tree trunks on slopes with a southern aspect may experience their maximum heat load in winter (Krenn 1933). The loading is increased above a snow cover and the large diurnal fluctuations of bark temperature may result in the splitting of the bark by frost.

2.2.5 The distribution of moisture near a bare surface

The fluctuations of moisture that occur in the air near the ground can be correlated with temperature. As the temperature increases the capacity of the air to hold water also increases; thus there is an increase in evaporation rate if water is available. Evaporation in response to increased temperature results in an increase in the partial pressure of water vapour (i.e. vapour pressure). Thus, during the day, maximum vapour pressures might be expected to occur at ground level, lower pressures obtaining in the upper air and resulting in the net movement of water away from the surface. However, during the day convection mixes the air and thus the maximum vapour pressure is likely to be found at the ground surface in the morning and afternoon when air movement is less (Figure 2.12). At night the minimum temperatures and vapour pressures are likely to occur at the surface and there will be a net flow of water vapour from the air down to the surface. If temperatures at the surface are sufficiently low, dew formation will result. A consideration of the heat budget makes it possible to predict dew formation by formula (Hofmann 1955) by a method not dissimilar to the Penman formula for evaporation (Penman 1948). Monteith (1963) discusses the physical basis and the ecological relevance of dew formation. The vertical temperature gradients set up in the soil at night also promote the upward movement of water vapour in the soil and some of this may be distilled as dew. This distillation represents a redistribution of existing soil moisture, which, in an experimental situation can provide a supply of moisture to plants in dry soil (Müller-Stoll & Lerch 1963). Dewfall from the atmosphere represents a net gain of moisture. However, the amount of moisture gained by condensation is only a small fraction of that added by precipitation or lost by evaporation. Geiger (1965) gives an estimate of 30–40 mm per annum in the German Federal Republic which can be compared with an annual precipitation of 771 mm and an estimated evaporation of 416 mm (Keller & Clodius 1956).

Relative humidity is the ratio, expressed as a percentage, of the amount of water held in the air to the amount that would be held at full saturation at the same temperature. The saturation vapour

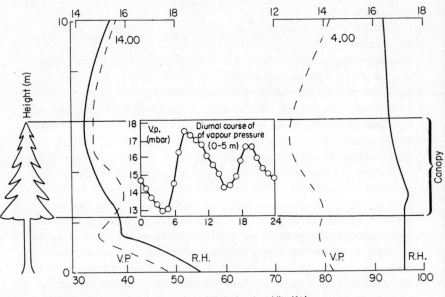

FIGURE 2.12. Vapour pressure and relative humidity in a stand of young spruce at the warmest part of the day and the coolest part of the night (derived from Baumgartner 1956).

Note that the periods of maximum and minimum vapour pressure (inset) are not coincident with temperature maxima and minima (cf. Figure 2.11 and Figure 2.10), and the capacity of the canopy to retain moisture.

pressure, and consequently the amount of water that can be held at full saturation, increases with temperature and thus relative humidity is dependent upon temperature. If the amount of water in the air remains constant then relative humidity will show an inverse relationship with temperature. Profiles of relative humidity above the ground surface tend to show such a relationship; thus both maximum and minimum relative humidities are found near the ground, the former before sunrise and the latter in the early afternoon (Figure 2.12). In a cool and moist climate the greatest humidities, by both day and night, are found at the ground surface, whilst in arid climates the reverse situation obtains and the lowest humidities are found at the ground surface (Geiger 1965).

2.2.6   The distribution of moisture within vegetation

During the day, when maximum temperatures are found near the surface of a vegetation canopy, the inverse correlation between temperature and relative humidity leads to a minimum of relative humidity within the canopy and a maximum near ground level (Figure 2.12). The lack of temperature stratification within dense stands results in little difference in relative humidity between the ground and the top of the canopy; however if any temperature inversion develops, then the maximum humidity will be found near the ground.

Maximum vapour pressures are found within the vegetation in the early morning and late afternoon when temperatures are relatively high and air movement is less pronounced (Figure 2.12). During the middle part of the day, air movement leads to reduction of vapour pressure within the canopy; this may be accentuated by reduced evaporation in dry periods when effective stomatal closure occurs. At night the lower temperatures are associated with lower vapour pressures.

The formation of dew, by condensation from the atmosphere, is likely to be associated with the height at which the greatest radiative cooling occurs. This point is in the vegetation canopy whereas it is near the surface of open ground (Figure 2.13).

The vegetation canopy also affects the distribution of rainfall; some water (*intercepted water*) is retained by the canopy and evaporated later, other water flows down twigs, branches and the trunk to reach the ground near the bole of the tree (*stemflow*) whilst other water penetrates directly or drips off leaves and branches (*through-fall*). The relative proportions of these components is not constant for any species of tree or configuration of stand (e.g. Ovington 1954) and is likely to vary with the intensity of the rainfall. However, it is possible to make a long-term approximation by assuming that the canopy has a certain capacity to absorb rainfall which is at first independent of the amount of rain and then shows some proportionality with rainfall (Horton 1919).

In winter the interception of snow is found to be less effective than that of rain. Deciduous trees lack leaves and less interception might be expected, but the same holds true for conifers. It appears that heavy wet snow falls through on account of its own accumulated weight and that dry powdery snow invades the canopy with ease.

A fuller account of the water relations of plants is given in Chapter 6 of this book.

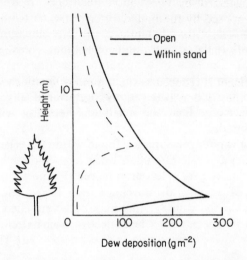

FIGURE 2.13. Dew deposition inside and outside a stand of young spruce (data of Baumgartner 1956).

2.2.7   *Air movement above the ground and within vegetation*

The air above a surface is never stationary. Wind is a horizontal mass movement of air, also known as advection, that is usually associated with gross differences in air pressure. However, even in the absence of a steady wind, 'pockets' of air move in all directions on account of convective movement. The resulting turbulence is largely responsible for the transfer of heat, water vapour, carbon dioxide and other gases, and suspended matter to and from the surface. Otherwise the transfer of heat would be solely by conduction and radiation and that of water vapour and other gases by diffusion. In the absence of air movement the migration of heat and water vapour from the surface would be almost equally limited, as the transfer coefficients are similar (0.25 cm^2 s^{-1} for water vapour; 0.22 cm^2 s^{-1} for heat), and diurnal fluctuations would not be detectable at any distance from the surface. In fact, diurnal temperature fluctuations can be detected up to 1 000 m above a warmed surface and local cloud formation during summer provides evidence that water vapour is transferred over large vertical distances (Geiger 1965).

There is, however, a narrow zone next to any surface which is entirely insulated from air movement. Within this 'boundary layer' heat is transferred by conduction and gases by diffusion. The resistance imposed by the boundary layer limits the movements of materials to and from the surface and consequently is of great importance in the heat economy and gaseous exchanges of leaves (3.1.2).

Air movement is reduced as the ground is approached, largely on account of the friction between air and ground (Figure 2.14). At

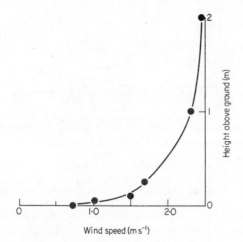

FIGURE 2.14 Variation in windspeed above a bare ground surface. (Dunblane, Perthshire 25.5.75: unpublished data. The assistance of Paul Bannister is gratefully acknowledged.)

low wind speeds this decrease allows the development of extremes of temperature and vapour pressure near the ground and consequently the profiles of temperature and water vapour often show a similar pattern to that of wind. High winds forcibly mix the air and phenomena such as dew deposition and ground frosts do not develop in windy nights. There is also a diurnal periodicity in air movement. The maximum degree of turbulence is found about noon when the large temperature differences near the ground cause differences in air density and encourage convection. The calmest periods are in the middle of the night (Figure 2.15).

The wind profile above a vegetation canopy is similar to that above the ground (Figure 2.16) so that there is comparatively little wind movement within vegetation. The maximum reduction of wind speed

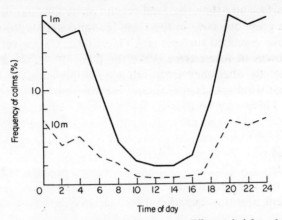

FIGURE 2.15. Annual incidence of calms at different heights above the ground surface (Berlin—data of Henning 1957).

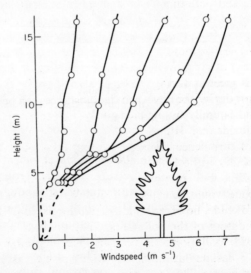

FIGURE 2.16. The variation of windspeed with height above and within a stand of spruce (after Baumgartner 1956).

The five curves plot the relationship at different windspeeds.

is near the surface of dense vegetation, such as an exposed coastal heath, but wind more readily penetrates flexible and vertically orientated vegetation such as grass. The effective surface for the reduction of wind speed is therefore somewhere between the actual ground surface and the top of the vegetation canopy and its height above the ground is used in formulae for the computation of wind speed above vegetated surfaces (e.g. Gates 1962, Geiger 1965).

The crowns of forest trees brake the force of the wind (Figure 2.16) especially when they are in full leaf (Geiger and Amann 1932) but at high wind speeds the canopy may deform and become less resistant. There may be increased wind movement between trunks and flows of cold air at night (katabatic wind) may be favoured by forest cover as their passageway between the ground and the crowns is insulated from external disturbance (Ungeheurer 1934).

The attenuation of air movement within vegetation aids the stability of temperature and moisture regimes and thus the environment beneath vegetation is less variable than that outside.

2.2.8 *The distribution of dust and gases near the ground*

Dust, smoke and polluting gases such as sulphur dioxide originate from the ground or near it. Carbon dioxide, apart from its output in smoke, results from the activity of organisms living in, on or near the ground. The distribution of all these substances is related to the diurnal pattern of air movement. Thus their content in the air near the ground is greatest during the night when there is little air movement and least during the day when the substances can be transported vertically and laterally by air movement (Figure 2.17). In the case of carbon dioxide (e.g. Huber 1952) it is not possible to disentangle the effects of turbulence from the diurnal cycle of CO_2 exchange shown by plants, as both show the same pattern.

At a constant level of pollution the distribution of smoke and sulphur dioxide would be expected to show similar patterns to dust and carbon dioxide, however their distribution is also related to the intermittent output of the pollutants. Thus the peak of smoke and sulphur dioxide pollution is in the early morning after fires are lit and there is a decline during the day which may be associated with increased turbulence. The level drops at night as fires are allowed to die (Meetham 1956). The development of temperature inversions above urban areas may effectively trap pollutants and a 'smog' may

be produced. The chemical constituents of smog may be exotic, for
example the interaction of petrol fumes and sunlight produces ozone
and peroxides in amounts which are certainly irritating and possibly
harmful to man. In normal situations ozone originates from the
upper atmosphere, and it is interesting to note that then the distribution of this gas follows the opposite pattern to that of a gas
emanating from near the ground, with a maximum near midday
when eddy diffusion brings down ozone from above and a minimum
at night when air movement is slight (Teichert 1955).

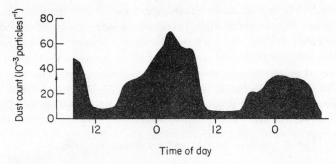

FIGURE 2.17. Dust count above the ground in a meadow in summer.
(Simplified and redrawn, after Effenberger 1940).
   Note the high dust content at night when the incidence of calmer periods
allows dust to settle out.

The lessened air movement within a stand of vegetation has a
marked effect on any substance contained in the air. The oscillations
in content that occur outside the vegetation are likely to be dampened
within the vegetation and heavier particles will tend to fall out of the
slower moving air. Vegetation reduces dust and condensation nuclei
(Zenker 1954).
   Pollutants may be concentrated by vegetation, the greatest
interception being by plants and plant organs with a large surface
to volume ratio (e.g. Gorham 1959). The continual accretion of litter
results in a further concentration: for example, Ovington &
Lawrence (1964) showed that the concentration of Strontium-90 in
oak litter was more than 25 times greater than that in the leafy
canopy.
   The planting of trees in urban areas with high pollution can be a
sanitary measure in that pollution will be intercepted. However,

such trees will have only a short life span and are unlikely to add to the aesthetic appeal of the landscape.

Vegetation also encourages the precipitation of fog. Studies of this interception are documented by Hori (1953) for the Japanese island of Hokkaido which is plagued by sea-fogs; similar fogs are caused by the interaction of moisture laden air and cold seas off the coasts of Eastern Britain and Newfoundland. Ooura (1953) on Hokkaido found that the fog was precipitated at the rate of $0 \cdot 5$ mm hr^{-1} with a moderate wind speed and Grunow (1955) has estimated that there may be an increment of more than 50 per cent to the annual precipitation by the interception of fog and cloud at moderate altitude (> 900 m) in Southern Bavaria. Under freezing temperatures the deposition may result in an accumulation of rime and its sheer weight may cause mechanical damage to trees, such damage is worsened if the top-heavy trees are subjected to wind.

This chapter has attempted to provide some physical background to the study of physiological ecology, with particular regard to the aerial environment. Subsequent chapters deal with the more detailed responses of plants to their environment.

3: Energy Balance and Gaseous Exchanges in Plants

3.1   THE RADIATION BALANCE AND
      DIFFUSIVE RESISTANCE OF LEAVES

3.1.1   *Radiation balance*

The shortwave radiation (I) incident upon a leaf is partly absorbed (a) but some is reflected (r) and transmitted (t). Thus

$$I = a + r + t \quad \text{(Figure 3.1)}$$

The component that is most important for the physiology of the leaf and hence the whole plant, is the amount absorbed; this is complementary to the amounts transmitted and reflected. All components are influenced by wavelength (cf. Figure 2.6) solar angle and the proportion of direct and indirect components in the incident light. The anatomy and morphology of the leaf has a profound influence; there are differences not only between the leaves

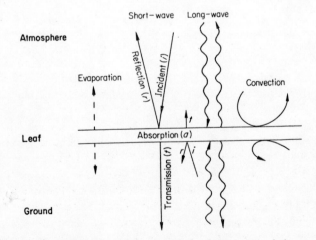

FIGURE 3.1. Diagrammatic representation of the energy balance of a horizontal isolated leaf.

of different species but also between leaves of the same species, even from the same plant. Values for a variety of species, some of which are referred to in the following text, are assembled in Table 3.1. Many plants (e.g. *Robinia pseudoacacia*) absorb less than 50 per cent of the incident shortwave radiation but others (e.g. *Pinus strobus*) may absorb almost all. Thick leaves will transmit less light than thin ones and thus the transmission of leaves of *Canna indica* is found to vary between 4 and 29 per cent with a mean of 10 per cent (Raschke 1956). Transmission may be greater through the upper surface (e.g. *Populus alba* where the upper surface is glabrous and the lower one hairy) but may also be greater from beneath (e.g. *Ulmus rubra*). Cuticularization, rugosity and hairiness all influence reflection from leaf surfaces and internal reflections also occur. Thus the felted lower surface of the white poplar (*Populus alba*) reflects almost twice as much light as the upper surface. Further values for reflectance may be found in published work (e.g. Billings & Morris 1951, Pearman 1966).

TABLE 3.1 Reflection, transmission and absorbtion of solar radiation by leaves. (values in %) (mainly derived from Birkebak & Birkebak 1964).

Species	Upper surface			Lower surface		
	r	t	a	r	t	a
Populus alba	28	18	54	44	18	38
P. tremuloides	32	19	49	36	19	45
Salix babylonica	28	22	52	31	20	49
Betula alba	30	22	48	33	24	43
Quercus rubra	27	24	49	29	25	46
Ulmus rubra	24	27	49	31	22	47
Prunus serotina	24	19	57	33	21	46
Acer platanoides	25	24	51	29	25	46
A. negundo	31	22	50	32	19	49
Robinia pseudoacacia	32	26	42	38	26	36
Fraxinus pennsylvanica	29	21	50	33	15	52
Alocasia (= *Canna*) *indica* (Raschke 1956)	21	10	69	—	—	—
Pinus strobus	11	0	89	—	—	—

Most calculations of the energy budgets of leaves are on the basis of a plane, isolated and horizontal leaf in air. In practice, leaves are variously contorted, show different orientations and exist in canopies where they receive not only incident radiation but also light transmitted through and reflected from other leaves. Although some leaves show a permanent horizontal orientation (e.g. rosette plants) many leaves are capable of altering their orientation with respect to a light source (cf. Huber 1935, Väupel 1958). Seedlings of *Acer rubrum* show horizontal orientation in low light intensities but the leaves become vertically oriented under high insolation (Grime, 1966). Similar movements may occur in seedlings of *Acer platanoides* and *A. pseudoplatanus* and are readily observed in the wood sorrel (*Oxalis acetosella*) in shaded and insolated habitats.

Converse forms of orientation have been recorded in arctic plants (Warren Wilson 1957) which may place their leaves perpendicular to the sun's rays and thus absorb the maximum amount of solar radiation. The cloudberry, *Rubus chamaemorus*, has relatively large leaves which are horizontally oriented in upland bogs and is so efficient at absorbing radiation that it has been described as 'warm to the touch' (Idle 1970). Many plants growing in low light intensity also show phototropic, responses which not only involve bending towards the light but also placing their leaves at right angles to the incident light. The longwave energy budget of leaves is simpler than for shortwave radiation. At normal temperatures leaves approximate to black bodies and thus both absorb and emit large amounts of longwave radiation. Usually about 95 per cent of the longwave radiation is absorbed (Gates & Tantraporn 1952).

The energy gained by the leaf is dissipated by convection and evaporation (transpiration) and the amounts used to heat the leaf and for photosynthesis are small in comparison, and can be neglected in considerations of gross energy balance. Thus the total radiation balance (S_L) can be considered as

$$S_L = LE_L + C_L$$

where L is the latent heat of evaporation, E_L the transpiration rate and C_L the convective exchange of heat. The balance between transpiration and convection depends on the relative resistances of the leaf to water vapour and heat transfer and is considered later (3.1.3).

The amounts used in photosynthesis and heating the leaves, although small, are of great significance for the plant. Some idea of their proportions under natural conditions can be gained from data for a stand of Norway spruce (Baumgartner 1956). The maximum solar input was about 800 Wm^{-2} whereas the maximum heat exchange was c. 35 Wm^{-2} (4 per cent) and the maximum amount used in photosynthesis only 10 Wm^{-2} (1 per cent). An isolated leaf would use more of the incoming radiation in photosynthesis than a complete canopy which only has its topmost leaves fully exposed but the amount used is still likely to be quite small.

3.1.2 *The diffusive resistance of leaves*

The resistance (R_L) of a leaf to the diffusion of gases is proportional to the concentration difference ($C_L - C_A$) that exists between leaf and air and is inversely proportional to the rate of flow (Q). Thus

$$R_L = (C_L - C_A)/Q.$$

This is analagous to Ohm's Law in electricity. The equation allows the diffusive resistance of a leaf to be calculated from a knowledge of concentrations and rates of flow. Its solution is simplest for water vapour as the internal leaf surface can be considered to be saturated and the water concentration of the ambient air can be estimated from measurements of relative humidity and temperature. Considerations of energy balance or direct measurements of transpiration can be used to calculate rates of flow.

The diffusive resistance to carbon dioxide is less readily calculated, for while it is possible to measure external concentrations and rates of assimilation of carbon dioxide, the internal concentration must vary with the rate of photosynthesis. When carbon dioxide is limiting photosynthesis, as at light saturation, internal concentrations will be at their minimum. This concentration is usually estimated by the carbon dioxide compensation point (4.2.2) and used in the equation.

The resistances within a leaf can be partitioned. There is an internal (mesophyll) resistance to transfer (r_m), an external resistance (comprising boundary layer resistance and resistances to eddy diffusion, r_a), a cuticular resistance (r_c) and a variable stomatal resistance (r_s). These last two resistances are in parallel, the rest in series (Figure 3.2).

$$R_L = r_a + r_m + \frac{r_s r_c}{r_s + r_c}$$

Cuticular resistance, particularly to the diffusion of carbon dioxide, is very large in comparison with stomatal resistance and its contribution can often be ignored. For example, if cuticular resistance is only 100 times greater than stomatal resistance then

$$\frac{r_s r_c}{r_s + r_c} = \frac{1 \times 100}{1 + 100} = \frac{100}{101} \simeq 1 = r_s.$$

However, cuticular transpiration is sometimes estimated as high as 20 per cent of transpiration with open stomata, in situations such as this cuticular resistance cannot be ignored although such high values may be partly due to incomplete stomatal closure.

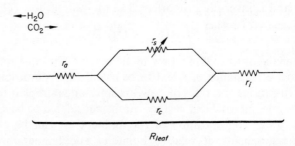

FIGURE 3.2. Electrical analogue of resistances in a leaf.
r_a leaf-air resistance, r_s stomatal resistance, r_c cuticular resistance, r_i internal resistance, R_{leaf} total leaf resistance.
Arrows indicate the general direction of movement of water and carbon dioxide through the system.

The external resistance r_a is often estimated by considering the evaporation rates of replicas of the leaf, usually made from blotting paper or plaster of Paris and saturated with water (e.g. Cowan & Milthorpe, 1968; Thom 1968). The resistance of other components may be estimated by considering the gaseous exchanges of leaves under strictly defined environmental conditions (e.g. Holmgren, Jarvis & Jarvis 1965). The mesophyll resistance to carbon dioxide transfer has to be extended to include components for aqueous transport from cell wall to chloroplast and a biochemical resistance which depends upon the rate of utilization of carbon dioxide within the cell.

Some workers (e.g. Meidner & Mansfield, 1968, Heath 1969) have used a theoretical approach based on the dimensions of diffusion pathways. Fick's Law is used for free diffusion and Stefan's Law for diffusion through pores. The external resistance (r_a) is the most difficult to estimate by these means as the model assumes a plane, circular leaf fully saturated with water and in perfectly still air. Dissected and irregularly shaped leaves are likely to have a lower r_a (Lewis 1972) and any air movement will also reduce r_a. Empirical corrections for nominally still air and moving air have been derived by Milthorpe (1961) and Penman & Schofield (1951). The basic equation relates leaf resistance (R_L) inversely to the

TABLE 3.2. Calculations of leaf resistances for a hypothetical leaf with the following characteristics:

effective diameter (d_L) 25 mm
leaf thickness (L) 0·2 mm
stomatal depth (L_s) 5 μm
stomatal diameter (d_s) 10 μm
stomatal frequency (n) 200 mm^{-2}
diffusion coefficient (water) (D) 0·249 $cm^2\ s^{-1}$

Calculations	Resistances (s cm^{-1})					
	r_i	r_a	r_s	R_{leaf}	rel. (%)	stom. (%)
Nominally still air	0·40	2·73	0·33	3·46	100	10
Absolutely still air	0·40	3·95	0·33	4·68	135	7
Moving air (10m s^{-1})	0·40	1·47	0·33	2·20	64	15
50% decrease in:						
leaf thickness	0·20	2·73	0·33	3·26	94	10
leaf diameter	0·40	1·98	0·33	2·71	78	12
stomatal frequency	0·40	2·73	0·67	3·80	110	18
stomatal depth	0·40	2·73	0·27	3·40	98	8
stomatal aperture	0·40	2·73	0·91	4·04	117	23
90% reduction in stomatal aperture	0·40	2·73	14·82	17·95	519	83

The relative resistance compares the total leaf resistance with that in nominally still air. The stomatal resistance is also expressed as a percentage of total leaf resistance.

The following formulae were used to compare resistance:

$$r_i = \tfrac{1}{2}L/D;\ r_s = \left(L_s + \frac{\pi d_s}{4}\right)/(na_sD) = \left[\frac{L_s}{a_s} + \frac{1}{d_s}\right]/D_n$$

where a_s = area of stomatal pore.

r_a (absolutely still air) $= \pi d_L/8D$

r_a (nominally still air) $= (\pi d_L^{0.6})/8D$

r_a (moving air) $= \left[\frac{0{\cdot}89\ d_L^{0.44}}{u^{0.56}}\right]/D$

where u = wind speed

$R_{leaf} = r_i + r_s + r_a$

diffusion coefficient for water or carbon dioxide and directly to the total effective length of the diffusion pathway. This latter component comprises an external (leaf-air) pathway, a stomatal pathway which is a function of the depth of the stomatal pore and inversely related to stomatal size and frequency, and an internal pathway within the leaf. The units of R_L are in s cm^{-1} or, if results are expressed on a whole-leaf basis (e.g. Meidner & Mansfield 1968, Heath 1969), in s cm^{-3} (see also 6.3.2).

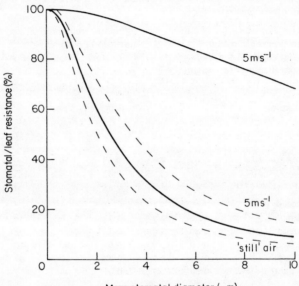

FIGURE 3.3. The relative contribution of stomatal resistance to total leaf resistance in transpiring and assimilating leaves at various windspeeds. Based on data of Heath (1969) for a hypothetical *Pelargonium* leaf.
————transpiration resistance (H$_2$O)
- - - - - -assimilation resistance (CO$_2$)
   The greater contribution of stomatal resistance to total leaf resistance, particularly evident in moving air, in transpiring leaves indicates that stomatal closure will be more effective in reducing water loss than in limiting CO$_2$ uptake.

These relationships provide a useful model that allows the quantification of alterations in leaf morphology and anatomy (Table 3.2). It is well known that leaf resistance to gaseous flow is increased by larger leaves, fewer, deeper and narrower stomata and thicker leaves but Table 3.2 shows that in leaves with open stomata changes in r_a are likely to have the greatest effect upon total leaf resistance and that when r_a decreases the relative contribution of the stomatal resistance increases. Small leaves therefore combine rapid gaseous exchange when the stomata are open with effective control when they close and are consequently suited to environments where favourable conditions are transient. Large leaves will show less rapid gaseous exchanges and a slower stomatal control and are suited to a more equable environment where their size will be an added advantage in the interception of radiation for photosynthesis.

The diffusion resistance to carbon dioxide will be greater. This is partly because the diffusion coefficient (D) for carbon dioxide is lower than that for water vapour but also because of the greater effective length of internal pathway. The diffusion coefficient for carbon dioxide in water is $1/10^4$ of that in air, thus an aqueous pathway across the cell of only 1 μm is equivalent to an aerial pathway of 1 cm. This increase in internal pathway plus any biochemical resistance makes stomatal resistance more important in the control of water movement than in the control of carbon dioxide assimilation (Figure 3.3) and represents an excellent ecological arrangement as it means that stomatal closure will not restrict photosynthesis to the same degree as it restricts water loss. This differential effect is even more marked in moving air (Figure 3.3).

3.1.3 *Leaf resistance and heat transfer*

The convective transfer of heat from a leaf (H) is related to the effective leaf diameter (d_L), the temperature difference between leaf and air $(T_a - T_L)$ and the wind-speed (u) as follows:

$$H = K(T_a - T_L) \cdot (u/d)^{\frac{1}{2}} \quad \text{(Gates 1962)}$$

where K is a constant depending on the morphology of the leaf.

The association with leaf diameter relates heat transfer to the leaf's external resistance to diffusion (r_a) and r_a for heat transfer is generally considered to be identical with that for water transfer. The equation also infers that nonevaporative heat transfer will be greater with large temperature differences, high wind speeds and small leaves.

Water loss is related to a total leaf resistance (R_L) and the gradient of water vapour $(C_L - C_a)$ between leaf and air and it is possible to determine the ratio (B) of convective (sensible) heat transfer and evaporative transfer by combining the appropriate equations.

$$\text{thus} \quad B = K \frac{(T_L - T_a)}{(C_L - C_a)} \times \frac{R_L}{r_a}$$

where K is the psychometric constant. This ratio (analogous to the Bowen ratio in meteorology) enables one to see that convective transfer will be proportionately high when there is a large temperature or small water vapour difference between leaf and air or when the total resistance of a leaf is high in relation to its external resistance (e.g. when stomata are closed).

(a) Genotypic differences

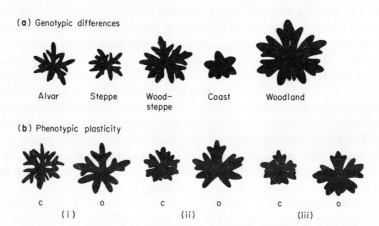

(b) Phenotypic plasticity

FIGURE 3.4. Genotypic and phenotypic differences in leaf size and dissection for representative leaves of *Geranium sanguineum* (adapted from Lewis 1972).
  (a) leaves from five populations from sites of differing aridity grown under uniform conditions.
  (b) leaves of representative clones (i–iii) grown under simulated continental (c) and oceanic (o) conditions.
Note the decrease in size and the increase in dissection of leaves from more arid and more continental conditions.
N.B. 'alvar' is an arid limestone community.

These various theoretical treatments allow certain ecological conclusions to be drawn. Large leaves have a high external resistance, this means that a considerable reduction in stomatal aperture will have to occur before transpiration is effectively controlled. When their stomata close they will not be very effective at dissipating heat by convection and will tend to overheat. On the other hand they have the advantage of a large photosynthetic area. Large leaves therefore seem to be adapted to moist, shaded areas and are generally found in such situations in nature. Small leaves will show a more rapid stomatal control of water loss and will also effectivèly dissipate heat when the stomata are closed and are thus adapted to dry, open, insolated habitats (Figure 3.4). Other morphological adaptations such as variations in leaf thickness, and hairiness, stomatal modifications and leaf dissection can be seen to combine some of the advantages of large leaves with those of smaller units. The various anatomical modifications will make internal resistances a larger proportion of the whole and thus make stomatal control more effective, leaf dissection will reduce external resistance and improve both stomatal efficiency and the dissipation of heat.

3.1.4   Theoretical models

The various leaf resistances can be used to compute photosynthesis as well as water loss. Photosynthesis is most readily considered at light saturation when only carbon dioxide is limiting:

$$\text{then } P = \frac{C_a - C_i}{R_{CO_2}}$$

where C_a and C_i are the concentrations of carbon dioxide in the external air and inside the leaf and R_{CO_2} is the resistance to the diffusion of carbon dioxide. This resistance differs from the resistance to water loss as it must include components for the acqueous diffusion of carbon dioxide across the cell and a biochemical resistance which will depend upon the rate of utilization of carbon dioxide at the chloroplast (Rackham 1966, Heath 1969).

More advanced models must take account of the interaction of water loss and photosynthesis with temperature, the relationship between heat transfer and leaf temperature and the effects of wind. The energy balance of the leaf must be computed and this involves a consideration of short- and longwave radiation balances and the convective and evaporative transfer of heat. The detail of the various equations and their solution is beyond the scope of this text but they are related to those included in the previous sections (3.1.1, 3.1.2, 3.1.3). The equations are not independent, for example, the temperature of a leaf is related to the heat loss by evaporation which is in turn determined by the gradient of water vapour concentration from the leaf to the air; but the concentration of water at the leaf surface is also determined by its temperature. Nevertheless, solutions are possible and examples of this sort of approach are shown by the work of Thom (1968), Cowan & Milthorpe (1968) and Lewis (1972) (Figure 3.5).

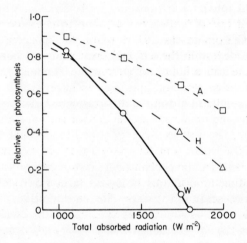

FIGURE 3.5. Computed photosynthetic responses of leaves of *Geranium sanguineum*, based on considerations of diffusive resistance, energy balance and a generalized relationship of photosynthesis to temperature. (Lewis 1972).

(A) Alvar leaf, (W) woodland leaf (see Figure 3.4), (H) hypothetical leaf with intermediate characteristics. Note that the smaller, more dissected, alvar leaf is estimated to be more effective than the woodland leaf, particularly when the radiation load is high.

3.2   STOMATAL MOVEMENT

Although energy balance and gaseous exchanges are influenced by leaf morphology, stomatal movement is of prime ecological importance as it regulates all gaseous exchanges—particularly those involving carbon dioxide and water vapour. The control of water loss has often been emphasized in ecological literature at the expense of the reduction in carbon dioxide assimilation that must also occur, but stomatal closure does not only restrict water loss, it also limits photosynthesis and prevents the evaporative cooling of the leaf surface.

The effects of light are manifest in the diurnal rhythm of opening and closing that is associated with light and dark periods. Stomata respond also to the quality of light and show more rapid responses in blue rather than red light. Temperature alters the speed of response and opening is more rapid at high temperatures (Meidner & Mansfield 1968).

In the last twenty-five years the role of carbon dioxide has been emphasized. Stomata close in response to high concentrations of carbon dioxide within the leaf and open when the concentration is low. Thus the utilization of carbon dioxide by photosynthesis during the day results in stomatal opening and its evolution by nighttime respiration results in closure. Temporary closure at midday can be explained by the abnormal production of carbon dioxide by over-heated plants as well as by a response to desiccation. Even the apparently abnormal stomatal opening that occurs at night in many succulent plants can be explained by the response to carbon dioxide as their acid metabolism (3.3.1) utilizes carbon dioxide in the build-up of organic acids in the dark. The reduction in carbon dioxide concentration allows stomatal opening whilst the breakdown of the acids during the day releases carbon dioxide and encourages closure (Figure 3.6).

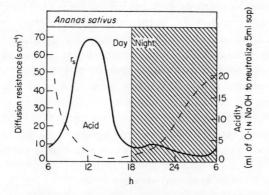

FIGURE 3.6. Stomatal movement in pineapple (a plant with crassulacean acid metabolism) in relation to light and leaf acidity (after Aubert 1971). High values of r_s indicate closure, lower values opening.

Stomatal response is thus the resultant of interactions between light, temperature and carbon dioxide concentration which lead to changes in the turgor of guard cells. Changes may, of course, also be induced by the direct loss of water from the guard cells. Light and low carbon dioxide concentrations lead to turgid guard cells and stomatal opening whilst the opposite conditions cause flaccidity and closure. The mechanism is often explained in simple osmotic terms with the photosynthesis and low acidity that occur in the light resulting in the production of osmotically active sugars whilst the acid conditions that occur in the dark favour the conversion of sugar to osmotically inactive starch. This is an oversimplification and Meidner & Mansfield (1968) discuss many of the current theories in greater detail. More recently (Raschke 1975, Meidner & Willmer 1975) emphasis has been laid upon the role of active transport of ions, particularly potassium, and an active response provides a better explanation of the rapid responses which may be shown by stomata.

There are not many comparative ecological studies of the responses of stomata to environmental factors other than desiccation. Recently the opening responses of the stomata of various trees to short periods of bright light have been examined (Woods & Turner 1971) and it was found that a shade tolerant beech (*Fagus grandifolia*) responded much more rapidly than the intolerant tulip tree (*Liriodendron tulipifera*) while trees of intermediate tolerance (*Acer rubrum, Quercus rubra*) showed intermediate responses (Figure 3.7). The stomata on opposite surfaces of the same leaf may also show different responses to light: in tobacco stomatal closure on the upper surface of a leaf is induced by a light intensity ten times higher than that required to induce closure on the lower surface of the same leaf (Turner 1970). Shade plants also show a more rapid stomatal closure in response to desiccation than do sun plants (Bannister 1971). Results such as these suggest that the stomata of plants or leaves from shaded habitats will respond more rapidly to increased light intensity, and be less inhibited by low light intensities, than those from more open habitats. This will enable carbon dioxide assimilation, and therefore photosynthesis, to respond to the fluctuating light intensities that occur in shade whilst the sensitivity to desiccation will prevent undue damage to the delicate leaves that are produced in shaded conditions.

In broad ecological terms, the responses of stomata to environmental factors ensure that stomata are open during the day when carbon dioxide is usually being utilized in photosynthesis, and closed during the night when carbon dioxide is produced. The nightly closure allows the amelioration of water deficits that have been induced during the previous day and some conservation of carbon dioxide for use the following day. The effects of blue light and temperature ensure that the most rapid responses occur in the middle of the day rather than in the less severe conditions of the morning and the evening. High temperatures are potentially deleterious as they could accelerate opening responses and expose the plant to desiccation on account of the rapid evaporation that is normally associated with high temperatures. However, the desiccation itself would rapidly lead to a loss of turgor in the guard cells and resultant stomatal closure. This response of stomata to excessive water loss is a valuable protective mechanism and is discussed elsewhere (6.43) but the principal adaptation of stomata is evidently to facilitate the gaseous exchanges of photosynthesis.

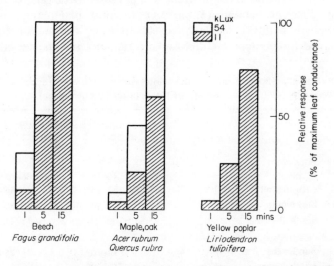

FIGURE 3.7. Stomatal responses of leaves from trees of varying shade tolerance to brief exposures of bright light (i.e. artificial sunflecks). Data of Woods & Turner 1971.
   The most rapid responses are shown by the most shade tolerant species (beech) and the least rapid by the least shade tolerant (yellow poplar) where there is no difference in the reaction to the two light intensities.

3.3 PHOTOSYNTHESIS, RESPIRATION AND CARBON ECONOMY

3.3.1 *Photosynthetic mechanisms*

The biochemistry of the photosynthetic process is well known and it is comprehensively treated in the many texts on the subject. Its treatment here is in a very simplified form, sufficient to draw attention to the ecological consequences of different processes. These are treated more fully in Chapter 4.

The light reactions appear to be qualitatively similar in all green plants although there is much ecological diversity in the responses of plants to the amount of light available (e.g. 4.3.2, 4.3.3). The light reactions activate chlorophyll molecules and, through phosphorylations, provide reducing power that is used for the production of carbohydrate from carbon dioxide as well as a supply of ATP. The various dark reactions are markedly sensitive to temperature and carbon dioxide concentration and such sensitivities probably account for some of the observed differences in photosynthetic rates of plants from different habitats (Table 3.3). In most plants the pentose phosphate pathway of Calvin cycle operates. Carbon dioxide is accepted by ribulose 1,5 diphosphate (RuDP) and the six carbon

TABLE 3.3 Maximal rates of net photosynthesis (on an area and a dry weight basis) in various groups of plants (after Larcher 1973a).

| | CO_2 assimilation | |
	mg dm^{-2} h^{-1}	mg g^{-1} (dw) h^{-1}
Herbaceous plants		
C_4 plants	50–80	60–140
Crop plants	20–40	30–60
Sun plants	20–50	30–80
Shade plants	4–20	15–25
Deciduous trees		
Sun leaves	10–20 (25)	15–25 (30)
Shade leaves	5–10	
Evergreen trees	4–15	3–18
Mosses	c. 3	2–4
Lichens	0·5–2	0·3–2 (3)
Algae	—	c. 3

compound which is formed splits to produce two molecules of phosphoglyceric acid each with three carbon atoms. (Hence this process is also known as the 3-carbon (C_3) pathway). Further reduction produces phosphoglyceraldehyde which is used as the basis for the synthesis of various carbohydrates from which the RuDP is eventually regenerated (Figure 3.8).

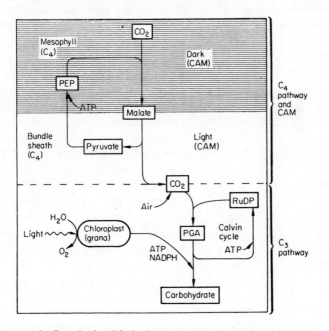

FIGURE 3.8. Greatly simplified scheme to show the relationships between 'normal' (C_3) photosynthesis, the Hatch-Slack pathway (C_4) and crassulacean acid metabolism (CAM).
   PEP phospho-enol-pyruvate; RuDP Ribulose diphosphate; PGA phosphoglyceric acid; ATP adenosine triphosphate; NADPH reduced nicotinamide adenine dinucleotide phosphate.

However, alternative processes also exist and are of ecological importance. The Hatch-Slack pathway (Figure 3.8) uses phosphoenolpyruvate (PEP) as an acceptor and has been found in various tropical grasses and cereals (e.g. sugarcane, maize and grainsorghum) and in some dicotyledonous species, notably members of the genera *Amaranthus* and *Atriplex*. A comprehensive check-list is given by Downton (1971). All plants possessing this pathway can reduce intracellular carbon dioxide to a very low concentration and continue to photosynthesize. Photosynthesis is enhanced when carbon dioxide is limiting as a steep carbon dioxide gradient can be maintained across the leaf. Thus, effective photosynthesis can occur when stomatal apertures are reduced due to water stress. The efficient re-utilization of respiratory carbon dioxide may partly account for the ability of many of these plants to show positive photosynthetic gains at high temperature (Figure 3.9) although the temperature sensitivity of the carboxylating enzyme systems also differ (e.g. Phillips & McWilliam 1971).

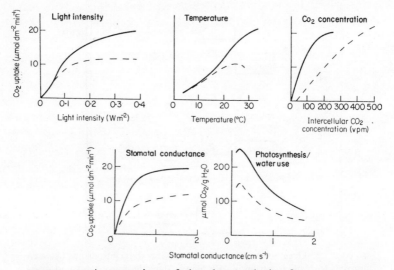

FIGURE 3.9. A comparison of the photosynthesis of two species of *Atriplex* with different photosynthetic mechanisms (based on Björkman & Berry 1973). ———— *A. rosea* (C_4) - - - - - - *A. patula* (C_3).

The Hatch-Slack pathway is associated with a distinctive anatomy where the mesophyll is comparable to that of normal plants but there is also a bundle sheath which produces much starch and has choroplasts with very poorly developed grana. The fixation of carbon dioxide by PEP occurs in the mesophyll and produces aspartate and malate which are transported to the bundle sheath. These acids have four carbon atoms and this route is also known as the 4-carbon (C_4) pathway. The acids are broken down to produce carbon dioxide which is used in the conventional Calvin cycle and pyruvate which is passed back to the mesophyll and used to regenerate PEP.

The spatial segregation of acid metabolism that occurs in C_4 plants is replaced by a temporal separation in a related process, crassulacean acid metabolism (CAM), that occurs in many succulents (e.g. members of the Crassulaceae, Cactaceae, Liliaceae, Bromeliaceae etc.). Here PEP is used to produce malate during the night (Figure 3.8). The plants maintain open stomata at night and take up carbon dioxide which is incorporated into malic and other organic acids with a resultant increase in the acidity of the cell contents. The organic acids are broken down in the light with the release of carbon dioxide, which is subsequently used in the Calvin cycle, and a corresponding decrease in intracellular acidity. This mechanism allows succulents to photosynthesize with closed stomata during the day and to assimilate carbon dioxide through open stomata during the night when water stress is less severe. However, when temperatures are cool or when there is little thermoperiodic fluctuation, stomatal opening and gaseous exchanges are at a maximum during the day; furthermore, the fixation of CO_2 at night is inhibited under long days (Ting 1971) and in times of severe moisture stress (Szarek & Ting 1974). It is therefore possible that C_3 metabolism prevails except during the specific conditions that allow CAM. The latter appears to be a survival mechanism rather than an efficient means of photosynthesis.

On the other hand, the high rate of carbon dioxide uptake shown by leaves of C_4 plants often leads to the conclusion that this pathway is 'superior' to the C_3 pathway. However, when growth rates of C_3 plants are compared with those of C_4 plants, the latter appear to have no inherent advantage. The growth rates shown by beans (C_3) are comparable with those of sugarcane and maize (C_4) and exceeded by those of sunflower (C_3) (Bull 1971). Growth rates appear to be determined more by the investment in leaf area than the rate of carbon dioxide assimilation. The high photosynthetic rates of C_4 plants may be characteristic only of young, unstressed tissues. In sugarcane they do not persist in mature or droughted leaves; these show similar maximum rates at high light intensity to C_3 plants and lower rates at low light intensities. The maximum growth rate of sugarcane was thus found to coincide with minimum photosynthetic rates (Bull 1971). It seems therefore that C_4 metabolism is an adaptation to dry regimes with high temperatures and light intensities. Both C_4 and CAM pathways appear to be modifications of the usual, C_3, metabolism that represent adaptations to specific environments: under normal conditions, C_3 plants are likely to be equally or more efficient.

3.3.2 Respiration in the light and in the dark

Dark respiration is the main energy-yielding process in organisms and its biochemistry is well known. Net photosynthesis is often considered to be the difference between dark respiration and gross photosynthesis, the main energy-fixing process.

However, both the utilization of oxygen and the evolution of carbon dioxide in the light appear to be greater than would be expected by dark respiration alone: indeed, it is uncertain whether dark respiration occurs to any great extent in illuminated leaves. Any 'photorespiration' is obscured in the light by the normal gaseous exchanges of photosynthesis. Evidence for the process is found in the transient 'burst' of carbon dioxide evolution that occurs when illuminated leaves are suddenly darkened and in the suppression of this burst at reduced oxygen tensions. Photorespiration uses oxygen in the production of glycollate and in its further oxidation to phosphoglycerate and serine: carbon dioxide is evolved during the breakdown of glycollate. The process consumes rather than yields energy and competes with the photosynthetic system for substrates and reducing power (Tolbert 1971). Consequently, it must result in a loss of photosynthetic efficiency. The phenomenon is most marked in plants with a normal (C_3) photosynthetic pathway; its apparent absence in plants with a C_4 metabolism is almost certainly due to their efficient re-utilization of any evolved carbon dioxide, although it is possible that the glycollate pathway is suppressed in some species. If photorespiration is inhibited by placing C_3 plants in atmospheres with low oxygen, then their rates of carbon dioxide assimilation may approach those of comparable C_4 plants (Björkman 1971).

The ecological significance of photorespiration has not been fully evaluated. However, as the gaseous exchanges of photorespiration are generally greater than those in the dark, gross photosynthesis is likely to be underestimated if it is based on the summation of net carbon dioxide assimilation and dark respiration.

3.3.3 Measurement of photosynthesis and respiration

The process of photosynthesis involves the utilization of radiant energy to combine carbon dioxide and water to produce carbohydrate, water and oxygen. Theoretically at least, photosynthesis could therefore be measured by estimating energy usage, carbon dioxide assimilation, the net assimilation of water or the production of carbohydrate or oxygen. Most common methods are based on the measurement of the assimilation of carbon dioxide (Bowman 1968) particularly as a sensitive measuring instrument, the infra-red gas analyser, exists. As carbon dioxide is also produced in respiration, most methods measure *net* (or apparent) photosynthesis rather than *gross* (or real) photosynthesis. Oxygen evolution may also be measured although measuring instruments are not normally as sensitive as the infra-red gas analyser. However, oxygen evolution is often used to measure the photosynthesis in aquatic systems where the oxygen produced in photosynthesis is a significant proportion of the total dissolved oxygen and can be readily detected by polarographic techniques (oxygen electrodes).

Relatively large amounts of energy, water and carbohydrate are stored or used in processes other than photosynthesis in the plant. This renders the small amounts used or produced in photosynthesis difficult to detect against the background of other usage. An estimate of the amount of energy used in photosynthesis involves precise measurements of the various uses of the energy absorbed by the leaf—much of this is used to heat the leaf or is dissipated by evaporation or convection—small errors in the measurement of these components can lead to large errors in any estimate of photosynthesis. Measurements based on the utilization of water in photosynthesis, although theoretically possible, are impractical as the water used is already contained within the plant and it would be very difficult to identify this water and separate it from that produced in photosynthesis or used elsewhere within the plant. The use of isotopically labelled water (H_2O^{18}) allows the estimation of the amount of isotopic oxygen evolved, but its measurement requires the use of a mass spectrometer—a facility that it is not available to most field ecologists. On the other hand, the use of labelled carbon dioxide ($C^{14}O_2$) is the basis of a field method, particularly for aquatic plants. Care has to be taken to ensure that the measured radioactivity in the samples emanates from compounds that have been produced in photosynthesis rather than from absorbed or dissolved carbon dioxide.

Most of the methods for measuring photosynthesis involve the enclosing of whole or part of the plant in some sort of chamber. The microclimate of a chamber rapidly diverges from that outside— light may be reduced or increased, temperature and moisture are likely to increase. Consequently, measurements of photosynthesis may be obtained which are not representative of those outside the chamber. The use of a continuous flow of air or water and sophisticated control of chamber microclimate (Lange *et al.* 1969) help to minimize such effects, but can prove expensive. For many ecologists the solution is to enclose the plant for only a very short time.

More detailed considerations of the methodology and problems involved in the measurements of photosynthesis are to be found elsewhere (e.g. Heath, 1969, Šesták *et al.* 1972, Chapman, 1976).

Respiration is usually measured by the evolution of carbon dioxide or the utilization or oxygen in the dark. The infra-red gas analyser is readily utilized to measure the former whilst classical manometric techniques, e.g. the Warburg apparatus, in which carbon dioxide is absorbed, measure the latter.

3.3.4 *Growth analysis*

As a plant grows it produces more photosynthetic tissue which then allows more photosynthesis. This leads to an exponential increase in growth (although the value of the exponent varies with the stage of growth) and thus it is more valid to examine growth in terms of logarithmic rather than arithmetic units.

The science of growth analysis evolved from observations on the growth of crop plants and has recently been considered in detail by Evans (1972). It essentially involves the examination of the rate of change of dry weight per unit plant weight with time and can be derived from fitted growth curves obtained from frequent small harvests (Hughes and Freeman 1967, Hunt & Parsons 1974). However, approximations of relative growth rate (RGR) are usually made by two harvests separated by a short interval.

$$\text{mean RGR} = \frac{\ln(w_2) - \ln(w_1)}{t_2 - t_1} = \bar{R} \text{ (Evans 1972)}$$

where w_1, w_2 and t_1, t_2 were weights and times at the first and second harvest respectively. The true relative growth rate is given by

$$\text{RGR} = \frac{1}{w} \cdot \frac{dw}{dt} = \mathbf{R} \text{ (Evans 1972)}$$

The analysis is usually taken further by expressing dry weight increment as a function of leaf area. This measure is known as net assimilation rate (NAR) or unit leaf rate (ULR).

$$\text{mean NAR} = \frac{w_2 - w_1}{t_2 - t_1} \times \frac{\ln(a_2) - \ln(a_1)}{a_2 - a_1} = \mathbf{\bar{E}} \text{ (Evans 1972)}$$

where a_1, a_2 are leaf areas at the first and second harvests. The instantaneous net assimilation rate is given by

$$\text{NAR} = \frac{dw}{dt} \cdot \frac{1}{a} = \mathbf{E} \text{ (Evans 1972)}$$

The formula assumes that measures of dry weight and leaf area are linearly related, this may only hold good for short time intervals. Net assimilation rates can be converted into relative growth rates by using the ratio of leaf area to dry weight (a/w). This is known as the leaf area ratio (LAR).

$$\text{RGR} = \text{NAR} \times \text{LAR} = \frac{dw}{dt} \cdot \frac{1}{a} \times \frac{a}{w} = \frac{dw}{dt} \cdot \frac{1}{w}$$

The leaf area ratio may be considered as the product of the specific leaf area (leaf area/leaf dry weight) and the leaf weight ratio (leaf dry weight/plant dry weight).

Growth may also be analysed by considering the distribution of dry weight throughout the plant at various stages in growth. The total weight at each harvest is divided into components such as roots, leaves, stems flowers and fruits (Figure 3.10). This enables the investigator to evaluate the 'strategic' investment of dry matter, e.g. in reproductive structures, and forms the basis of the concept of 'reproductive strategy' (e.g. Harper & Ogden 1970).

Growth rates vary both with stage of growth and cultural conditions. Young plants and those grown under optimal conditions show high rates of growth and net assimilation. Some generalizations can be made on the basis of maximum rates of growth, either measured instantaneously or over a period ($\mathbf{R}_{max}$ of Evans 1972).

Crop plants often show higher rates than wild plants, woody plants show lower rates than herbaceous plants and broad-leaved deciduous trees higher values than evergreen conifers (Table 3.4). The low growth rates of woody plants and other perennials are probably related to their production of woody and storage tissue at the expense of leaves.

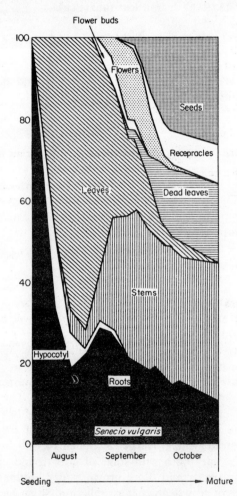

FIGURE 3.10. Allocation of dry weight to different structures through the life cycle of *Senecio vulgaris* (Harper & Ogden 1970: from Chapman 1975).

TABLE 3.4. Examples of extreme values of maximum relative growth and net assimilation rates of various species grown under favourable cultural conditions.

RELATIVE GROWTH RATE (g g^{-1} week $^{-1}$)	Low		High	Source
Broad-leaved trees				
Fagus sylvatica (5 years old)	0·05	*Populus tremula*	0·92	Jarvis & Jarvis 1964
Coniferous trees				
Pinus sylvestris (5 years old)	0·02	*Pinus sylvestris* (1 year old)	0·50	Jarvis & Jarvis 1964
Shrubs and dwarf shrubs				
Vaccinium vitis-idaea	0·23	*Salix cinerea*	1·06	Grime & Hunt 1975
Herbaceous crop plants				
Trifolium subterraneum	0·91	*Lycopersicum esculentum*	2.80	Jarvis & Jarvis 1964
Herbaceous wild plants				
Lathyrus montanus	$\begin{cases}0·46 \\ 0·49^*\end{cases}$	*Stellaria media*	$\begin{cases}2·09 \\ 2·43^*\end{cases}$	Grime & Hunt 1975
Grasses & Sedges etc.				
Schoenus nigricans	0·15	*Poa annua*	$\begin{cases}1·74 \\ 2·70^*\end{cases}$	Boatman 1972 Grime & Hunt 1975

NET ASSIMILATION RATES mg cm^{-2} week^{-1})				
Broad-leaved trees				
Sour orange	1·8	apple grafts	6·9	Jarvis & Jarvis 1964
Coniferous trees				
Pinus sylvestris (5 years old)	1·1	*Pinus sylvestris* (1 year old)	3·6	Jarvis & Jarvis 1964
Herbaceous crops				
Hordeum vulgare	6·8	*Zea mays*	15·2	Jarvis & Jarvis 1964
Wild plants				
Impatiens parviflora (old plant)	1·0	*Holcus lanatus*	10·0	Coombe 1966 Hunt & Parsons 1974.

N.B. The values referred to in Jarvis & Jarvis (1964) originate from a variety of sources and are quoted more fully in that paper.

*Asterisked values of relative growth rate are instaneous maxima (R_{max}), otherwise they are average rates over a period.

A recent investigation of R_{max} for a wide range of species grown under standard conditions (Grime & Hunt 1975) has allowed the tentative identification of certain strategies. High R_{max} is associated with productive habitats where either the rapid occupation of available space is essential for competitive success (e.g. *Urtica dioica* with

$R_{max} = 2 \cdot 35$) or rapid growth to maturity is an advantage, as in weeds (e.g. *Poa annua* with $R_{max} = 2 \cdot 70$). Unproductive habitats are associated with plants of low R_{max} (e.g. *Sieglingia decumbens*, $R_{max} = 0 \cdot 60$) as this appears to confer resistance to environmental stresses and allows the plants to make low demands upon limited resources.

3.3.5 *Carbohydrate content*

Growth analysis examines the investment of photosynthate as dry weight. The incorporation of material as soluble sugars and reserve carbohydrate represents readily available resources which can be used as a 'current account' on a day to day basis. Periods of net photosynthesis are represented by carbohydrate gain whilst losses occur during periods of intense metabolic activity or when photosynthesis is absent. Translocation within the plant can result in gains or losses in particular organs.

Figure 3.11 shows the seasonal fluctuations in carbohydrate in a deciduous dwarf shrub. Carbohydrate is lost during the winter when the shrub is leafless and this loss may be related to the prevailing winter temperatures (Figure 3.12). Winter losses are correspondingly less in plants which can photosynthesize during winter. A rapid decline in carbohydrate content occurs during leaf expansion but this is soon compensated for by the photosynthesis of the new leaves. There is a decline in carbohydrate content during the period of vegetative growth that occurs in late summer but an increase around leaf fall when carbohydrate is transported from the leaves into the remainder of the plant. Soluble carbohydrate shows a maximum during winter months and a minimum during summer, this is correlated with the annual course of frost resistance (Figure 4.23). Observations of the respiration of deciduous trees (e.g. Schulze 1970) are inversely correlated with these observations of carbohydrate content during the growing season. Respiration is high in the spring and autumn and is associated with the growth that occurs during these periods.

Carbohydrate analyses can be used as the basis for comparisons between plants. Plants well adapted to a particular environment would be expected to show higher levels of carbohydrate throughout the annual cycle of growth than those under some stress (e.g. Russell 1940).

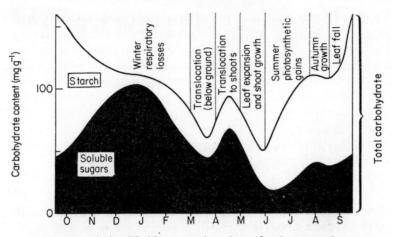

FIGURE 3.11. A simplified interpretative scheme for the seasonal course of carbohydrate content in above-ground shoots of *Vaccinium uliginosum*. (Based on observations of Stewart & Bannister 1973).

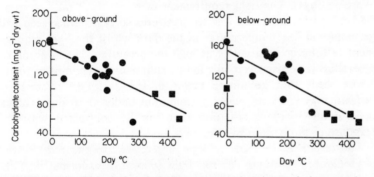

FIGURE 3.12. The total carbohydrate content in winter of above- and below-ground; parts of *Vaccinium uliginosum* in relation to accumulated temperatures above a base of 4°C. ● 1968/69 winter; ■ 1969/70 winter. (Stewart & Bannister 1973).

4: Responses of Plants to Light and Temperature

4.1.1 *Light and germination*

Some seeds germinate better in the light than in the dark whilst others show a converse response. A brief exposure to light suffices for the seed of Grand Rapids lettuce, but the necessary period may be longer, as in *Epilobium* (Isikawa & Yokohama 1967) or birch seeds which show their best germination under long days (Black & Wareing 1955). The light requirement of many seeds may be over-ridden by both high and low temperatures so that lettuce seed will germinate at low temperatures in the dark whilst the light require-ment of tobacco seed is lost at high temperatures. The addition of giberellins may also overcome light requirements (Thompson 1969). Those seeds that germinate better in the dark, e.g. *Nemophila insignis*, may show improved germination under short days (Black & Wareing 1960)—a situation that can be contrasted with the response of birch seeds to long days.

The responses of seeds to light are related to the red: far-red phytochrome reaction. The red light (660 nm) in normal daylight induces the production of a phytochrome (P_{fr}) sensitive to far-red light (730 nm). The reversion to the red-sensitive form (P_r) requires more energy and is thus a slower process. Consequently, P_{fr} accumu-lates and eventually allows germination to proceed. Prolonged exposure to far-red light may inhibit germination and even overcome the effects of chilling (Black 1969); blue light may also be in-hibitory. Effects such as these may account for the poor germination of some species in the light.

There is proportionately more far-red light under vegetation and this may account for the inhibition of germination in many species. Lettuce germination is inhibited by light filtered through lime leaves (Black 1969) and the germination of *Chenopodium* species is inhibited in light with a red/far-red ratio comparable to that of shade (Cumming 1963).

Chilling and light requirements may be reduced in older, after-ripened, seed and responses to light may be important only for newly shed seed. Seeds that require light will be unable to germinate if buried (Wesson & Wareing 1967) whilst those with a dark requirement would need to be covered. Seeds with a light requirement will also be prevented from germinating if they fall into dense vegetation whilst the inhibitory effect of prolonged far-red irradiation may prevent seeds without a light requirement from germinating under similar circumstances. Thus, one ecological effect of light requirements is to prevent germination under unfavourable conditions. However, light requirement may also ensure a reserve of seeds that germinates intermittently over a period of time. Seeds of *Chenopodium botrys* show optimum germination at low temperatures under short days but at high temperatures during long days (Cumming 1963). This is an adaptation that would allow germination to proceed both in summer and in winter—a useful strategy for a weed.

There would, however, appear to be no unique strategy for a particular environment. Studies of desert plants (Evenari 1965, Koller 1969) show that seeds both with and without light requirements are adapted to different ecological niches.

4.1.2   *Temperature and germination*

Temperature has two major influences upon germination. The first is a pretreatment which allows dormancy to be broken; the second as a direct influence upon the rate and amount of germination. Both these influences are of ecological importance.

Seeds with low temperature requirements for the breaking of dormancy are likely to be from populations adapted to long cold winters; thus *Rubus chamaemorus* (cloudberry) has a boreal-circumpolar distribution and the seeds require at least five months of continuous low temperature stratification at 4–5°C before any germination is detected and up to nine months stratification is required for substantial germination (Taylor 1971). In contrast *Erica cinerea*, a plant with a markedly oceanic and western distribution is inhibited by low temperature pretreatment (Bannister 1965). The most effective temperatures for low temperature pretreatment are, contrary to popular belief, above freezing between 0–6°C (Eckerson

1913). The requirement for a low temperature pretreatment can be regarded as a mechanism that prevents germination occurring in midwinter. If seeds were moved to a colder climate, the low temperature requirement would be met early in the winter and encourage premature germination whereas a warmer climate would prevent seed germination as the low temperature requirements would not be met and, furthermore, short periods of higher temperature (c. 20°C) would nullify the effects of the previous cold treatment. The low temperature requirement of older seed is lessened, this ensures some germination even if the low temperature requirement is not met initially.

Some species are stimulated to germinate by a brief exposure (< 1 minute) to high temperatures. Seeds of *Calluna vulgaris* (Whittaker & Gimingham 1962) and *Erica cinerea* (Bannister 1965) are stimulated in this manner and both species are from vegetation that is managed by fire. The germination of *Calluna* is only slightly increased but that of *Erica* is increased from 3 per cent to 40 per cent and this may account for the temporary dominance of *Erica cinerea* after fire (Gimingham 1949). Germination is stimulated by fire in other vegetation, such as chaparral, although the effect is often caused by the charring of otherwise impermeable seed coats (Sweeney 1956).

The rate and amount of germination is influenced by temperature and a batch of seeds typically shows a minimum, maximum and optimum temperature for germination. Within a species there may be much variation, as in *Rumex crispus* where the germination of seeds at constant temperature in the dark reveals differences between plants and even locations within the same inflorescence (Cavers & Harper 1964). Such differences appear to be related to the size and development of the seed and may be of advantage as they ensure that all seeds do not germinate at once. However, these differences are minimized when seeds are grown under alternating temperatures and many species show a similar response with optimum germination under alternating rather than constant temperatures.

Thus, *Calluna* seeds germinate better when the temperature is alternated between 20°C (16 hr) and 30°C (8 hr) than at either temperature by itself (Gimingham 1960) and *Erica cinerea* shows a better germination when temperatures are alternated between 2°C and 30°C (Bannister 1965). Germination in the dark is also enhanced by alternating temperatures. Thompson (1970a) has evolved a method of characterizing the germination of a species in response to a range of constant temperatures. He plotted, for successive days, the maximum and minimum temperatures at which 50 per cent of the final germination had been reached (Figure 4.1). The method

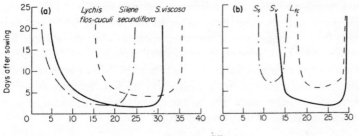

Temperature (°C)

FIGURE 4.1. Germination responses of caryophyllaceous species of different geographical origin in relation to temperature (abstracted and redrawn from Thompson 1970a).

Responses of old seed (a) and freshly-harvested seed (b).

Seeds are germinated on an electrically generated thermal gradient and the response expressed as the temperatures (maximum and minimum) producing 50 per cent germination at successive intervals after sowing.

Lychis flos-cuculi is characteristic of north-west Europe (France (a), England (b)).

Silene secundiflora is a Mediterranean plant (S. Spain).

S. viscosa is continental from a steppe climate (Hungary).

clearly indicates the effective maximum and minimum and the optimum range of temperatures for germination. The rate of germination is indicated by the nearness of the base of the curve to the ordinate. Rates of germination can also be compared by an Arrhenius plot where the logarithm of rate is plotted against the reciprocal of temperature. Thompson was able to show that caryophyllaceous species from the Mediterranean zone germinated over a low temperature range, species from the deciduous forest zone over a high temperature range whilst those from a steppe climate germinated over a wide range of temperatures. The responses are most marked in seeds which have been freshly harvested (Figure 4.1b) but are still demonstrable in older seeds (Figure 4.1a). The Mediterranean species are adapted to germination in the low temperature of winter, when moisture is freely available; deciduous forest species respond to the higher temperature of the summer which is the optimal period for germination in this zone; whereas the wide range of temperatures characteristic of exposed steppe and the availability of water in late summer is reflected in the response to species from this habitat. The minimum temperature for germination is generally lowered under an alternating temperature regime and the maximum may be raised although it appears to remain fixed in Mediterranean species. There are differences between the temperature responses of seeds from populations of the same species (Thompson 1970b) but, some widespread weeds (e.g. corncockle, *Agrostemma githago*) have retained the characteristics of their original Mediterranean habitat and are still adapted to germination at low temperature. Their survival has often been due to their seeds being unwittingly stored along with those of a cultivated crop.

4.1.3 Light, temperature and seedling establishment

The impact of light and temperature on the physiology of the mature plant is dealt with subsequently (4.3, 4.4), but the seedling may fail to survive to maturity and thus it is essential to examine the influence of light and temperature on the establishment of seedlings. Grime (1966) has considered the conflict between the adaptations which allow a plant to exploit a particular habitat and its need for survival as a seedling.

Most plants of dry, unproductive habitats are small although they may grow and mature rapidly during favourable periods. They produce many small seeds which facilitate rapid recolonization during moister periods and which do not require large amounts of energy for their production. Plants that produce small seeds which grow vigorously can rapidly colonize moist and productive habitats which have been temporarily cleared. Examples include weeds, such as thistles and willow herbs, and trees like birch and willow. The establishment of seedlings becomes difficult once a cover of vegetation has developed and seedlings must rapidly outgrow the shade of other plants. Larger seeds store enough food to allow seedlings to grow independently of photosynthesis (Figure 4.2a) and are characteristic of species from closed habitats.

Seedling establishment is a rare occurrence in most woodlands and the majority of herbaceous plants reproduce vegetatively; reserves of stored food allow them to grow rapidly early in the year before the leaf canopy is complete.

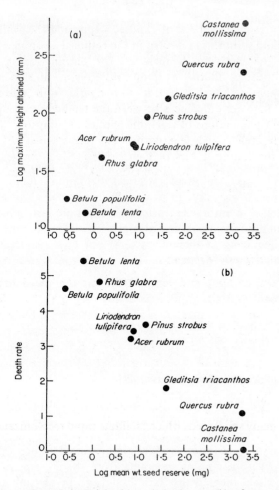

FIGURE 4.2. Growth in height (a) and death rate (b) of tree seedlings grown in artificial shade in relation to seed size (Grime & Jeffrey 1965).

In dense woodland it is essential that established tree seedlings survive until a permanent gap appears in the canopy. Large seed size, low growth rate and low respiration rates all help to conserve carbohydrate; seedlings that exhaust their reserves may become susceptible to fungal attack (Figure 4.2b) and those that show rapid elongation may also produce thin-walled cells which are easily penetrated by pathogens.

Temperature is likely to affect seedling establishment by influencing the balance between photosynthesis and respiration.
Respiration is more sensitive to temperature and seedlings commonly
have high respiration rates and little photosynthetic tissue. Respiratory losses will be most critical in seedlings which have become
established from small seeds with minimal food reserves.

Seedlings are likely to be subjected to extremes of temperature
in open habitats, particularly where the ground is dark-coloured, as
in recently burnt vegetation or in slag heaps. Sweeney (1956) records
differences in soil surface temperatures of more than 17°C between
burnt and vegetated areas; whereas Schramm (1966) refers to
temperatures of greater than 50°C, sufficient to cause heat-girdling of
woody stems, in anthracite waste. Bare areas are also subject to the
lowest minimum temperatures at night. These temperature extremes
may be damaging in themselves (4.4) but high temperatures will
additionally cause desiccation and upset the water balance of
seedlings.

4.2   ENVIRONMENTAL FACTORS
AND PHOTOSYNTHESIS

4.2.1   *Temperature and photosynthesis*

The effects of temperature upon plants and plant processes are not
simple although many interpretations of these effects adopt a simple
viewpoint. The process of photosynthesis is often considered to have
an optimum, maximum and minimum temperature (Table 4.1;
Figure 4.3) a view point stemming from the concept of 'cardinal
points' (Pisek 1960) which is a basic one in physiological plant
ecology and can be traced back at least to Sachs (1860). The action
of temperature is quite different for each cardinal point. The minimum
is coincident with the freezing of the plant tissues (Pisek *et al.* 1967)
whilst the maximum lies several degrees (2–12°C) below the thermal
death point and represents a balance between the inhibition of the
metabolic processes involved in photosynthesis and the increase of
respiration with temperature (Pisek *et al.* 1968). The optimum is
subject to limitation by factors other than temperature and thus the
optimum temperature for photosynthesis is raised as light intensity
increases (Figure 4.3); similar shifts of the optimum are to be expected

TABLE 4.1. Maximum, minimum and optimum temperatures for photosynthesis (net) in various groups of plants (Based on Larcher 1973a).

	minimum	maximum	optimum(°C)
Herbaceous plants			
Tropical (including C_4)	5–7	50–60	35–45
Crop plants (C_3)	−2–>0	40–50	20–30(40)
Sun plants	−2–0	40–50	20–30
Shade plants	−2–0	c. 40	10–20
Arctic-alpine	−7––2	30–40	10–20
Woody plants			
Tropical trees	0–5	45–50	25–30
Arid shrubs	−5––1	42–55	15–35
Temperate deciduous trees	−3––1	40–45	15–25
Evergreen trees	−5––3	35–42	10–25
Arctic/alpine dwarf shrubs	c. −3	40–55	15–25
Lichens (cold regions)	(−25)–15––10	20–30	5–15

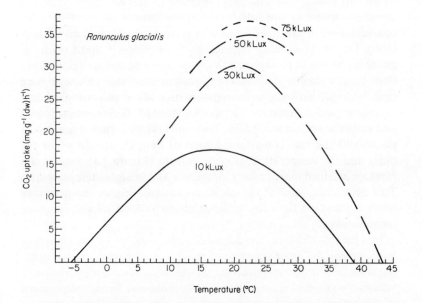

FIGURE 4.3. Net assimilation of CO_2 by *Ranunculus glacialis* in relation to temperature at different light intensities (after Pisek *et al.* 1969).
Note the higher optimum temperature at higher light intensities.

for increasing carbon dioxide concentration, but these are less ecologically relevant as atmospheric concentrations of carbon dioxide fluctuate between narrow limits. If an absolute temperature optimum exists, it can only be determined in the absence of any limitation by other factors, and this leads to a distinction between a physiological and an ecological optimum. A physiological optimum occurs when the plant is presented with optimum levels of light intensity, carbon dioxide and water content whilst the ecological optimum would be related to the overall influence of these factors in an optimal growing season. The ecological optimum is consequently shifted towards lower temperatures.

The rates of purely chemical reactions increase exponentially with temperature but metabolic processes are rapidly inhibited by high temperatures and the degree of inhibition increases with the duration of exposure, due to the progressive destruction of enzyme systems. Rapid measurements may record higher rates of reaction than those involving longer time intervals as the enzyme systems are more likely to be still intact. Consequently, short-term measurements shift the apparent optimum towards higher temperatures. A more detailed consideration of this phenomenon is given by Jost (1906) and Heath (1969, Ch. 7). Thus rapid measurements of photosynthesis made by gas analysis would be expected to give a higher optimum temperature than measurements derived from growth analysis; although long time intervals between measurements may allow physiological and morphological adaptation to occur (Mooney & Shropshire 1968) and raise the optimum slightly. The time taken to reach a maximum photosynthetic rate is also a function of both the condition of the plant and the temperature of measurement (Figure 4.4)—maximum rates are attained more rapidly in summer and at higher temperatures. This also tends to shift the optimum towards higher temperatures when a standard time for acclimatization is used over a range of temperatures.

In view of all these considerations the concept of an optimum temperature for photosynthesis (or, for that matter, many other physiological processes) must be treated with reservation. However valuable ecological information can be obtained from comparisons of the optimal temperature ranges of species from different climatic zones (Table 4.1). Thus *Acacia craspedocarpa* from semi-arid bush-land in Western Australia shows an apparent optimum at about 36°C, *Ficus retusa* from rainforest, an optimum at about 30°C;

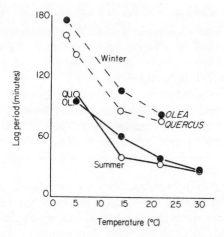

FIGURE 4.4. The time taken by leaves of *Olea europa* and *Quercus ilex* to reach maximum CO_2 assimilation in summer and winter (after Larcher 1969).
 The response is generally faster in summer and at higher temperatures than in winter or at lower temperatures.

Citrus limon from the Mediterranean region, an optimum of about 25°C; *Fagus sylvatica* in Central Europe 18°C whilst *Oxyria digyna* at 2 500 m shows an optimum at only 12°C. However, the temperature at the optimum for photosynthesis is not a species characteristic. Species from different habitats show different optima. *Betula pendula* from 1 900 m shows an optimum at c. 14°C whilst material of the same species from 600 m shows an optimum at 17°C; desert specimens of *Artemisa tridentata* show an optimum at about 25°C as opposed to c. 20°C in plants of the same species from a subalpine environment.

4.2.2 Compensation points

Low light intensities and both low and high temperatures may produce situations where apparent photosynthesis is zero because the respiratory output and photosynthetic intake of carbon dioxide are identical. These are compensation points which have the added advantage that they can be easily measured and readily interpreted in an ecological context.

 Compensation points are readily determined because they represent a point of no change. The use of bicarbonate indicators (e.g. Leith

1950, Martin & Pigott 1965) that are in equilibrium with the atmosphere provides a simple method for their determination. If a plant part gives off or absorbs carbon dioxide, the colour of the indicator will change; the compensation point is indicated by a point where the colour remains the same.

Other factors should be held constant during the measurement of compensation points otherwise they will change its value. Under normal atmospheric concentrations of carbon dioxide, temperature compensation points are influenced by light intensities and light compensation points by temperature (cf. Figure 4.5).

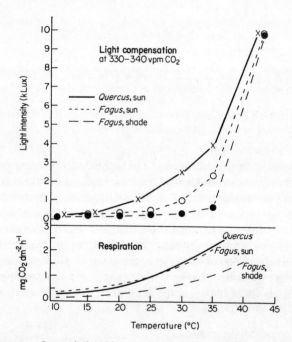

FIGURE 4.5. Interrelationships between light compensation point, temperature and respiration rate in *Quercus ilex* (sun leaves) and *Fagus sylvatica* (sun and shade leaves). (After Larcher 1969).

Sometimes the minimum concentration of carbon dioxide that can be produced in a closed system (Γ, (gamma) of Heath 1969) is referred to as a carbon dioxide compensation point, as when this point is attained there will be no net change in carbon dioxide concentration. The measurement of this point necessitates a precise measurement of carbon dioxide concentration and an infra-red gas analyser is commonly used (e.g. Meidner 1962). The value obtained is strongly influenced by both light and temperature, but its usefulness in providing a physiological parameter to be used in ecological comparisons has yet to be established. However, it seems that those plants (e.g. maize and sugar cane) which are capable of reducing the internal carbon dioxide concentration of their leaves to a very low level are also capable of high rates of photosynthesis (3.3.1). A low internal concentration will cause a steep gradient for the diffusion of carbon dioxide into the leaf, and as carbon dioxide is often a limiting factor in photosynthesis, will result in a heightened photosynthetic efficiency even when stomata are partially closed. Plants with a very low carbon dioxide compensation point typically incorporate carbon dioxide through the Hatch-Slack pathway (3.3.1) and are usually from dry, tropical habitats where water and carbon dioxide rather than light intensity, limit photosynthesis.

Low temperature compensation points

There is considerable variation in the resistance of plants to low temperatures (4.4) but the minimum temperature for photosynthesis is remarkably constant. Net assimilation ceases at the onset of freezing, whilst the complete freezing of free water within the tissues results in the absolute cessation of photosynthesis. Most plants show maximum rates of photosynthesis in summer when frost resistance is low. Alpine plants such as *Ranunculus glacialis* and *Oxyria digyna* show a minimum summer temperature for net assimilation as low as $-6°C$ which is correlated with an ability to sustain supercooling. Mediterranean species such as *Citrus limon* may show no net assimilation below a temperature of only $-1°C$, but the relationship between the low temperature compensation point and origin is inconstant. *Geum reptans* from the alpine zone ceases net assimilation at only $-3°C$, whereas the Mediterranean species, *Prunus laurocerasus* shows a low temperature compensation point at $-4°C$. In winter the minimum temperatures for net assimilation are somewhat lower (although nowhere near the frost resistance levels) and differences between species from different regions appear to be minimized (Table 4.1).

As the minimum temperature for net assimilation shows so little variation, it is possible that the effect of low temperatures on subsequent respiration and assimilation is of greater importance. Such effects have been demonstrated in plants of different growth form and origin (e.g. *Olea europea* (Larcher 1969), *Leucojum vernum* (Pisek *et al.* 1967) and *Abies alba* (Pisek and Kemnitzer 1968). In silver fir (*Abies alba*) the effects are greater in autumn than in winter and photosynthesis is more affected than is respiration (Figure 4.6). At even lower temperatures ($-14°C$) in winter the net assimilation may take several days to return to its original value, whereas when frost damage occurs the original rate is not reattained at all.

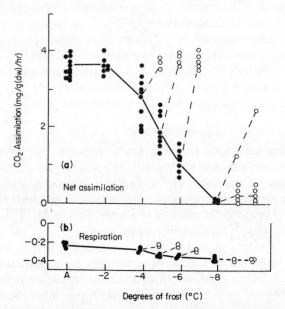

FIGURE 4.6. The effects of exposure (12 hours) to various degrees of frost upon the photosynthesis and respiration of silver fir (*Abies alba*) (Pisek & Kemnitzer 1968).
    Bold lines and filled circles—CO_2 exchange the morning after treatment (at $15°C$).
    Dashes and open circles—CO_2 exchange 24 hours later ($15°C$).
    A—unfrosted control.
    (a) CO_2 assimilation in the light, (b) respiration in the dark.

When leaves are cooled to temperatures above their freezing point, or when super-cooling occurs, recovery of photosynthesis may be immediate (Figure 4.7).

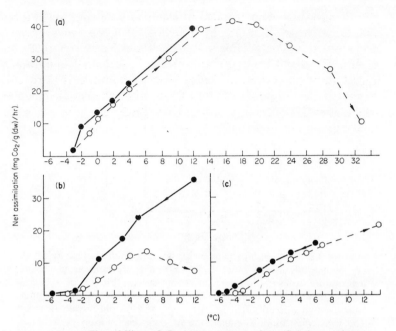

FIGURE 4.7. The ability of *Leucojum vernum* (snowflake) to photo-synthesize after being subjected to various degrees of frost (Pisek *et al.* 1967). Continuous lines and filled circles indicate cooling, dashed lines and open circles rewarming.
    (a) Cooled to −3°C, plant unfrozen. Recovery complete.
    (b) Cooled to −6°C, plant frozen and damaged, poor recovery.
    (c) Cooled to −6°C, plant frozen but undamaged. Recovery complete.

High temperature compensation points

Whilst the inhibition of photosynthesis at low temperatures is associated with the freezing of the assimilatory organs, at high temperature there is a more complex interaction. Net photosynthesis represents a balance between respiration (or photorespiration) and gross photosynthesis. As temperature rises there will be a greater output of respiratory carbon dioxide which may be partially re-assimilated at higher light intensities. Thus, the temperature at compensation is increased as the light intensity is increased. At constant light intensity, the intersection of curves showing the relationship of respiration and gross photosynthesis to temperature (Figure 4.8) must coincide with the high temperature compensation point.

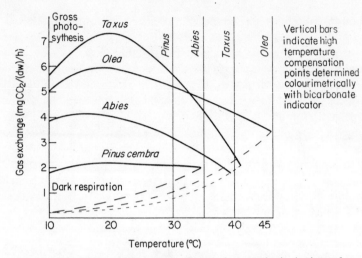

FIGURE 4.8. Relationship between gross photosynthesis (estimated as net photosynthesis + dark respiration), dark respiration and high temperature compensation point in various species (Pisek *et al.* 1968).

Note that in *Pinus cembra* the low compensation temperature results from a combination of high respiration rate and low photosynthetic rate. The high compensation temperature of *Olea* results from the relative insensitivity of photosynthesis to the detrimental effects of high temperature and a moderate respiration rate.

The discrepancies between values obtained by the direct measurement of gaseous exchanges and those obtained colourimetrically may be due partly to the longer equilibrium time necessary for the latter method causing a greater inhibition of photosynthesis.

High temperature compensation points (Table 4.1) appear to discriminate between species of different origins and ecological affinities rather better than do low temperature points. Species from high altitude (e.g. *Oxyria digyna, Ranunculus glacialis, Geum reptans, Pinus cembra*) show compensation at relatively low temperatures (36–39°C), whilst species of a general northern and continental distribution show a range between 37°C and 44°C (with coniferous trees showing the lower values and broad-leaved trees and shrubs the higher values). The highest range (42–48°C) is shown by species from the Mediterranean region.

Light compensation point

The light compensation point is the light intensity (typically measured in metre candles or lux) at which there is no net movement of carbon dioxide in or out of the plant leaf. At atmospheric concentrations of carbon dioxide the values of the light compensation point are strongly influenced by temperature; this effect may be very marked at high temperatures (Figure 4.5) and even at lower temperatures a change from 15–25°C raises the light compensation point of lettuce from 150 to 400 lux (Heath & Meidner 1967). The rise of compensation point with temperature appears to be correlated with changes in respiration rate rather than in photosynthetic rate as the latter is less sensitive to temperature. Accordingly, leaves with high respiration rates would be expected to show high compensation points and low compensation points would be expected from leaves with low respiration rates. Sun leaves have high compensation points which may be related to both the large amount of non-photosynthetic (and therefore wholly respiring) tissue also to the rapid extinction of light in a thick leaf with dense tissues. Young leaves have high respiration rates (Schulze 1970) and high compensation points, whilst shade leaves often have low respiration rates (Grime 1966) and almost inevitably have lower compensation points than sun leaves (Table 4.2).

The ecological importance of this type of adaptation is obvious. Plants or plant parts with low compensation points are able to make

TABLE 4.2. Light compensation and light saturation in various groups of plants (based on Larcher 1973a).

| | k lux | |
	compensation	saturation
Herbaceous plants		
C₄ plants	1–3	> 80
crop plants (C₃)	1–2	30–80
sun plants	1–2	50–80
shade plants	0·2–0·5	5–10
Deciduous trees		
sun leaves	1–1·5	25–50
shade leaves	0·3–0·6	10–15
Evergreen trees		
sun leaves	0·5–1·5	20–50
shade leaves	0·1–0·3	5–10
Mosses	0·4–2	10–20
Algae	—	(7)15–20

the maximum use of low light intensities. However, in many instances
the reduction of light intensity within a canopy is so severe that
self-pruning occurs. Parts of the plant that are below a critical light
intensity die and eventually fall off the tree. The critical light intensity will exceed the value of the compensation point as the
respiratory losses of a plant growing in the field are influenced not
only by the light intensity during the day but also by the duration
and temperature of the night. The incidence of self-pruning has
been correlated with the reduction of light intensity beneath the
canopies of different trees and shrubs (Wiesner 1907).

4.2.3 Light saturation

When all other factors are operating at optimum levels, photosynthesis increases as light intensity increases. However, the normal
atmospheric concentration of carbon dioxide is relatively constant
at around 300 ppm. and limits photosynthesis. Most land plants
show an increase of photosynthesis over a limited range of light
intensities until a further increase of light intensity results in little
or no further change in photosynthesis (Figure 4.9). The leaf is then
considered to be light-saturated and only an increase in carbon
dioxide above the normal atmospheric concentration will allow
enhanced photosynthesis. The abrupt cut-off is a consequence of the
high diffusive resistance of most leaves; in conditions where diffusive
resistance is low the carbon dioxide is rarely limiting and there is no

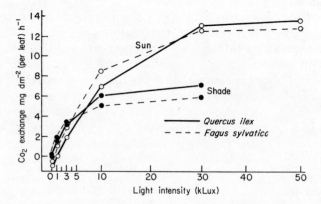

FIGURE 4.9. Light responses of sun and shade leaves of *Quercus ilex*
(sclerophyllous) and *Fagus sylvatica* (deciduous). (After Larcher 1969).

definite light saturation. Thus, water plants supplied with bicarbonate ions (Harder 1921), which effectively supply carbon dioxide directly to the reaction sites without setting up diffusion gradients, and C_4, plants which utilize carbon dioxide very efficiently, show no definite light saturation. The normal range of temperatures has very little effect upon light saturation (Larcher 1969).

A precise measurement of carbon dioxide assimilation over a range of light intensities is needed if the light saturation of a plant or plant organ is to be determined. It is difficult to define the exact point of saturation as saturation is likely to be achieved less definitely in a leaf with wide open stomata than in one with partially closed stomata. Consequently, light saturation is likely to be more definite in leaves attached to plants than in leaves with their stalks in water (which are likely to have wide stomata). Considerations such as these may explain the different results sometimes obtained for the same species. It is also difficult to decide the precise onset of light saturation because the curve approximates a rectangular hyperbola (Figure 4.9) with a continuous change in rate, and also because interpolation between measured points may also be necessary. However generalizations are usually possible and Figure 4.9 shows that shade leaves of both *Quercus ilex* and *Fagus sylvatica* are saturated at a light intensity of less than 10 kilolux whereas a value of somewhere between 10 and 30 kilolux is needed to saturate sun leaves of the same species.

Shade leaves generally show lower saturation light intensities than do sun leaves (Table 4.2), and the light intensity at saturation may also differ in geographical races of the same species. Alpine populations of *Oxyria digyna* show a higher saturation light intensity than arctic populations (Mooney and Billings 1961) and this difference, like that between sun and shade leaves, seems to be an adaptation to the prevailing light intensities. When light is reduced, those plants with a low light intensity at saturation will show a proportionately smaller reduction in photosynthesis than plants with higher saturation intensities. Consequently C_4 plants show a large percentage reduction in photosynthesis in poor light; this characteristic may help to confine these otherwise photosynthetically very efficient plants to open sunny habitats.

Most measurements of light saturation are made with isolated leaves—it is important to realize the difference between light saturation for the whole plant and that for only one of its leaves.

The whole plant exists in a vertical gradient of light and the light intensity below a canopy of leaves may be only a fraction (often less than 5 per cent) of the incident. Thus, despite the adaptation of leaves to shade, lower leaves are unlikely to be near light saturation and may well be below their compensation point. Consequently the light intensity for the maximum photosynthesis or dry matter production of a whole plant (or stand of vegetation) will be above that for an individual leaf. It has been suggested that the leaf area index required for the maximum production of dry matter is that which allows most shaded leaves to be brought above their compensation point; moreover, if levels of other factors are adequate, light is always likely to limit the production of natural communities (Blackman & Black 1959). Thus, as leaf areas index increases, the total production of a stand increases to a maximum but the production of individual plants declines (Figure 4.10).

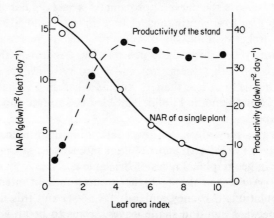

FIGURE 4.10. The relationships between the net assimilation rate of maize plants, stand productivity and leaf area index (from Larcher 1973a).

Oversaturation by light may be inhibitory (e.g. Jarvis 1964) but is often difficult to separate the effect of light from the increased temperature and imbalanced water relations that must also occur. Destruction of chlorophyll may occur in leaves subjected to high insolation and sun leaves generally appear more yellowish than shade leaves, but at high light intensities the amount of chlorophyll has a negligible effect upon the rate of photosynthesis and even at low light intensity (1·5 kilolux) it is only pale green or yellowish leaves that show a reduction in photosynthetic efficiency (Figure 4.11).

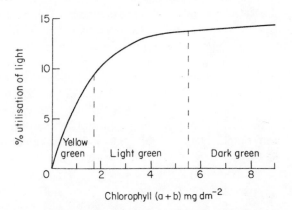

FIGURE 4.11. The relationship between light utilization and chlorophyll content of leaves (after Gabrielsen 1948—data for a number of species).

4.2.4   Morphological and physiological compensation for the effects of light and temperature within species

The preceding sections suggest that large differences in plant production and performance are to be expected in plants grown under different light and temperature regimes. However, differences are not always as great as might be expected.

At low light intensities, plants show physiological compensation and possess low compensation points and higher chlorophyll contents and are saturated by lower light intensities than those grown in full light, but otherwise light is seen as limiting plant growth. However, growth analyses of plants grown under different light regimes may show little differences between the rates of growth in the various regimes. The explanation for this observation is that the decline in net assimilation rate is compensated for by an increase in specific leaf area, i.e. leaves of shade plants have a greater leaf area per unit dry weight than those from sun plants. This morphological compensation is often found in plants which frequent shade such as the lesser celandine, *Ranunculus ficaria* (Figure 4.12), the bluebell, *Endymion non-scriptus* and the small balsam (*Impatiens parviflora*) and enables each species to show only small changes in growth rate over a wide range of light intensities (Blackman & Rutter 1948, Hughes 1965).

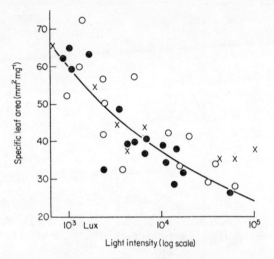

FIGURE 4.12. Specific leaf area of *Ranunculus ficaria* as a function of the light intensity at the site of collection (student data).
○ 1972, ● 1973, × 1975, each point is the mean of at least three leaves. The fitted curve is the form $y = \dfrac{\log x}{b} - a$ where $y=$ s.l.a and $x=$ light intensity and b ($= 268\cdot84$) and a ($= 28\cdot56$) are constants.

Physiological adaptation to temperature also occurs. Plants of the same species from different habitats show different temperature optima for photosynthesis (4.2.1) and plants kept at high temperatures generally show low respiration rates and thus conserve carbohydrate. Respiration rates can be increased by keeping the same plants at low temperature for as little as a week (Rook 1969). Plants from high latitudes and altitudes, where environmental temperatures are generally colder, usually show higher respiration rates than those from lower altitudes and latitudes (Figure 4.13). Such increases in respiration rate may represent a general adjustment of metabolic rate that enables plants to function effectively at low temperatures.

Light and temperature cause many more modifications within a species than those mentioned: such changes have been catalogued elsewhere (e.g. Daubenmire 1974). Many of the modifications are associated with changes in specific leaf area (e.g. changes in leaf area and thickness, and in cell size and stomatal frequency) and their significance is dealt with in other parts of this book (e.g. 3.1.2, 6.3).

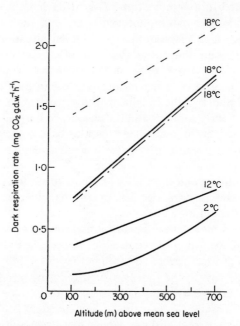

FIGURE 4.13. Relationships between dark respiration rate and altitude of origin in *Vaccinium uliginosum* (------), *V. myrtillus* (———) and *V. vitis-idaea* (—.—.—) (Stewart & Bannister 1974).

4.2.5 *Ecological aspects*

The balance between photosynthesis and respiration and its relationship to light and temperature can be invoked to explain many differences between plants of different ecological affinities. Compensation points provide an example of this kind of interaction that has already been considered (4.2.2).

The influence of respiration is critical, particularly when photosynthesis is limited by environmental conditions. Then, the factors affecting respiration become dominant and the carbohydrate reserves of the plant can become depleted if respiration exceeds photosynthesis for long enough. Depletion of reserves may lead to a lack of competitive vigour, failure to flower or set seed and lay the plant open to damage by extremes of environment or pathogens; whilst extreme exhaustion of the reserves could lead directly to the death of the plant. The failure of glasshouse and household plants in winter is often attributed to a combination of short daylength and low light intensity, leading to poor photosynthesis, and high temperatures which produce high respiration rates. Similar situations can occur under natural conditions, a combination of cloud, shortening days and relatively high temperatures at the arctic treeline in August has been shown to produce a net respiratory loss in pine trees (Figure 4.14). A net photosynthetic gain in these trees is confined to the longer days of early summer and the clearer, colder, days of autumn (Ungerson & Scherdin 1968). This tenuous balance between photosynthesis and respiration may provide a physiological explanation for the occurrence of an arctic treeline.

Altitudinal limits and the ability of plants to survive in shade may also be, at least in part, decided by the balance between photosynthesis and respiration.

Other factors may determine altitudinal limits. At high altitudes, the perpetual exposure to wind, rain, snow, sleet, hail and low temperatures (both at night and through the winter) may cause direct or indirect damage to the tissues, whilst the generally low environmental temperatures will inhibit growth, flowering, seed set and the maturation of tissues. Many plants of high altitudes adopt a resistant growth form, such as a cushion, which minimizes the effects of adverse environment, but which also renders these plants susceptible to competition from the more vigorous plants which can survive at lower altitudes.

In higher latitudes the altitudinal limits are lowered; this can be associated with increasing severity of the climate. However, at the same latitude, the altitudinal limits may differ. In Scotland the upper limit for potential forest is substantially lower in the west (90–135 m), where the climate is mild, than in the east which has a more severe winter climate (Pears 1968) and the altitudinal limit in the east (690–750 m) may be considerably lower than in continental Scandinavia (1 000 m). The same trend is shown by other types of vegetation (Figure 4.15). An explanation for this phenomenon may be found in the balance between photosynthesis and respiration. In oceanic climates, rates of photosynthesis are limited by low environmental temperatures and the general cloudiness of higher altitudes during

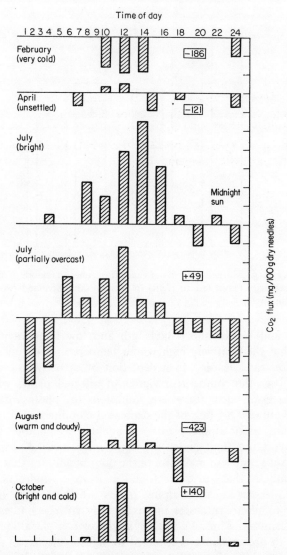

FIGURE 4.14. Photosynthetic economy of *Pinus sylvestris* on the arctic treeline (data of Ungerson & Scherdin 1968).

Figures in boxes on the right hand side of the diagrams indicate the mean daily loss or gain of CO₂ at the appropriate time of year.

Bars above each line represent net CO₂ gain, below net loss. Each division on the vertical scale represents the loss or gain of 10 mg CO₂ per 100 g dry needles.

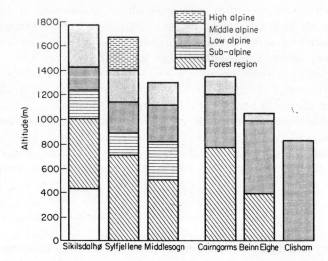

FIGURE 4.15. Altitudinal limits of vegetation zones in regions of increasing oceanicity (from left to right) in Norway and Scotland (Poore & McVean 1956).

summer and by the short daylength and low light intensities of winter, but the relatively high winter temperatures encourage high respiration rates leading to a depletion of carbohydrate reserves. This reduces the competitive vigour of affected plants which are forced to lower altitudes where conditions for photosynthesis are more favourable because of the decreased cloudiness. On the other hand many arctic-alpine plants are adapted to photosynthesis at low temperatures and under low light intensities (cf. Mooney & Billings 1961) and can make use of the poor conditions that exist for photosynthesis that are to be found at lower altitudes in oceanic climates. However, arctic-alpine plants which do not have the capacity to photosynthesize in winter may still be limited to the higher altitudes. Thus *Vaccinium uliginosum*, a deciduous dwarf shrub with brown stems, is confined to higher altitudes in the West of Scotland whilst *V. myrtillus*, which is also deciduous but has green stems, and *V. vitis-idaea*, which is evergreen, may also be found at lower altitudes (Stewart & Bannister 1973).

These considerations help to explain the peculiar assemblages of plants which may be found in parts of the West of Scotland and particularly in the Burren in the West of Ireland (Ivimey-Cook &

Proctor 1966) where evergreen arctic-alpine species such as *Dryas octopetala* and *Arctostaphylos uva-ursi* are found in the same area as species of a marked southern distribution such as the orchid *Neotinea intacta* and the maidenhair fern, *Adiantum capillus-veneris*. The evergreen arctic-alpines are able to photosynthesize at the low temperatures and light intensities that prevail whilst the southern species can occur because of the lack of winter cold.

Under a more continental climate the warmer and less cloudy summers allow more photosynthesis and the colder winters minimize respiratory losses. This enables larger species (e.g. trees) to survive at higher altitudes and thus competition confines the arctic-alpine plants to even higher altitudes. The occurrences of arctic-alpine plants at lower altitudes are often in areas where the competition from more vigorous species is excluded by peculiar site conditions, such as soils derived from calcareous or serpentinic rock with low fertility and high concentrations of naturally occuring heavy metals (Jeffrey & Pigott 1973).

The discussion outlined above is oversimplified because there is usually a range of phenotypic and genetic adaptations to local conditions within a plant species (4.2.4).

For example, in the case of respiration rates, plants with a wide altitudinal range might be expected to show more adaptation than plants with a limited range. This appears to be the case in *Vaccinium* species, where *V. myrtillus* and *V. vitis-idaea*, which have a large altitudinal range, show a greater adaptation of dark respiration rates than *V. uliginosum* which is generally confined to high altitudes (Figure 4.13).

Various aspects of plant adaptation to the shaded environment have already been considered (4.2.2, 4.2.3, 4.2.4). Photosynthesis in reduced light intensities is enhanced by the larger leaves, higher chlorophyll content, lower compensation and light saturation points of plants in shade. Respiration is reduced, largely through a reduction in the proportion of nonphotosynthetic tissue, and may also be less sensitive to increases in temperature (Grime 1966). All these factors contribute to a favourable balance between photosynthesis and respiration in plants adapted to the shade.

Light compensation points are essentially a measure of the balance between photosynthesis and respiration; species might be characterized by the change of their light compensation point with temperature (Figure 4.5). Species with a steep curve would be expected to show an unfavourable photosynthetic balance when subjected to changed light or temperature while those with shallow curves would be expected to be little affected by changes of environment.

4.3   LIGHT, TEMPERATURE AND DEVELOPMENT

The development and timing of many plant functions (phenology) is strongly influenced by both temperature and the length of the day, or photoperiod. In temperate climates and at higher latitudes both temperature and photoperiod alter over an annual cycle, for, as days lengthen, the positive net radiation balance results in higher temperatures and the converse occurs with shortening days. In tropical climates there is little annual variation and the biggest alterations in light and temperature are on a diurnal basis although a seasonal alternation of wet and dry periods may impose an annual periodicity.

Light and temperature may determine both the onset and cessation of growth as well as the rate of growth during the vegetative period. The cessation of growth is brought about by the build up of inhibitors (e.g. abscisic acid) in response to shortening days and falling temperatures so that maximum dormancy is developed by midwinter. Subsequently, rising temperatures and lengthening days result in dominance of stimulatory substances (e.g. giberellins) and dormancy is broken and shoot extension proceeds. These influences are mediated by the phytochrome system and it is therefore probable that the length of the dark period is more critical than the actual photoperiod.

Many annual plants, and particularly arable weeds, show a lack of periodicity in their responses and merely grow when environmental conditions are favourable. Common weeds such as *Senecio vulgaris*, *Cardamine hirsuta*, *Capsella bursa-pastoris* and *Arabidopsis thaliana* grow perceptibly during mild spells in the winter as well as during the summer. Perennial plants often show a more regular annual cycle of events which is influenced by environmental factors but has also an inherent (endogenous) component. Thus many tropical woody plants show a periodicity in their growth which may or may not be linked with fluctuations in the environment (e.g. Daubenmire 1972). Woody plants from the temperate regions almost inevitably show a periodicity of growth which can be correlated with environmental factors.

This interrelationship is well illustrated by heather, *Calluna vulgaris*. In common with many temperate plants, heather responds to long days by extension growth and flowering (Chouard 1946). Growth can be initiated by placing the plant, out of season, in artificially lengthened days. Extension growth is retarded during the flowering period, but if the plant is kept in continually long days it will enter a fresh phase of shoot extension followed by further flowering. In the field this is prevented by the shorter days and lowered temperatures that prevail towards the end of the growing season. In short days growth is slight and flowers do not form. However, growth can be initiated during short days if the plant is subjected to low temperatures and then placed in conditions suitable for growth. Plants kept outside during the winter will grow, under short days, when brought into the greenhouse in contrast with those brought inside in early autumn (Figure 4.16b). Growth can also be initiated under short days by artificially subjecting plants to low temperatures (Figure 4.16a). Thus both temperature and daylength interact in the initiation of growth. Many plants have a specific chilling requirement that must be satisfied before growth can commence.

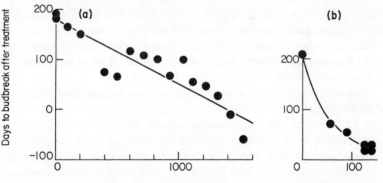

FIGURE 4.16. Modification of the daylength responses in *Calluna vulgaris* by prolonged exposure to low temperatures (unpublished data).

(The negative values in (a) occur where budbreak occurred in plants that were still being subjected to cold pretreatment. After pretreatment all plants were kept under short days under greenhouse conditions; in (a) plants were subjected to low temperatures at night and kept (under short days) in the greenhouse during the day.)

After initiation, subsequent growth is largely determined by temperature. Many temperate crops are considered to grow when daily mean temperatures rise above 6°C but a lower limit of 0°C may be more appropriate for plants of a more northerly distribution or from higher altitudes whereas species from warmer climates may require temperatures of 10–15°C for growth. Many species show adaptation to local temperatures and this is also true of heather. When grown in a common environment, populations from high altitudes flower earlier than those from lower altitudes (Grant & Hunter 1962) whilst populations from higher latitudes both flower earlier and show a more rapid shoot extension than those from lower latitudes (Bannister, unpublished data).

Adaptation to photoperiod also occurs. The duration of the juvenile, leafy, stage in gorse (*Ulex europaeus*) is inversely correlated with photoperiod and populations from northern Britain show an inherently shorter juvenile period (Milliner 1962).

Some plants (e.g. tobacco) have a requirement for short days. These also grow, often luxuriantly, in long days but fail to flower if they have had no pretreatment of short days. Their ability to grow under short winter days renders them susceptible to frost and to low temperatures which inhibit their growth, but they are well adapted to the mild moist winters of the Mediterranean climate where growth in long days may be inhibited by the typically dry summers. In Britain, the winter annual, *Hordeum murinum*, is limited to areas where the winters are sufficiently warm to allow growth and its distribution can be correlated with the presence of milder, urban microclimates (Davison 1970).

The adaptation of populations to local climatic conditions ensures that the individual is unlikely to develop at an unfavourable time, but on the other hand will limit its geographical spread. A long-day plant will have a southern limit set by the daylength required to initiate new growth and flowering and the failure to meet any chilling requirement. Northern limits would be set by the premature stimulation of growth, as chilling and daylength requirements would be met early in the season when there is still a risk of frost damage, and by low environmental temperatures during the growing season which causes the plant to be insufficiently 'hardened' by the end of the growing season.

Short-day plants can persist in more northerly latitudes, particularly if they have an efficient means of vegetative reproduction.

However such plants would be induced to grow during mild spells during the winter and would be rendered suceptibie to subsequent damage by frost and desiccation.

Altitudinal limits are set by prevailing temperatures. Plants with a small chilling requirement will be confined to warmer areas at low altitude (or nearer the sea) where they will not be induced to develop prematurely and temperatures during the growing season will be adequately high. Plants from high altitudes could have a large chilling requirement that would not be met at lower altitude and thus their development would be delayed, whilst their general adaptation to low temperatures might cause them to complete their development early in the growing season and lose the advantage of the longer growing season. Mild winter temperatures could also favour depletion of carbohydrate reserves (3.3.5). Temperature has an analagous influence on latitudinal limits.

In natural conditions the various selection pressures lead to the formation of populations adapted to local climates. The absence of such adaptation is of more concern to the agriculturalist or horticulturalist who might wish to extend the range of a particular species or variety. In some cases it is advantageous to grow a species out of its normal range as in the Maryland Mammoth tobacco, a short-day plant that produces huge leaves and no flowers in the longer photoperiods of a more northerly summer. There are also disadvantages —a northern limit is set to crops such as spinach and beet by their tendency to 'bolt' (i.e. produce flowers in one season rather than two). This condition is brought about by low temperatures early in the growing season and exacerbated by the long days of summer.

Trees and shrubs, whether grown for ornament or fruit, are often protected from the competition which would prevent their being successful in the wild but often fall foul of the local climate which frequently causes them to develop too early or grow on too late and renders them susceptible to spring and autumn frosts.

4.4   RESISTANCE TO TEMPERATURE EXTREMES

Temperature extremes, apart from limiting physiological processes (e.g. 4.2.1), may cause the death of whole or part of the plant and thus either completely eliminate it from a particular niche or reduce its competitive vigour. Consequently, a study of the resistance of plants to temperature extremes provides insights into their ecology. Investigations are usually made by subjecting whole plants or isolated parts of plants to the appropriate temperatures (see Bannister 1976).

Resistance to temperature extremes is conveniently subdivided into resistance to heat and cold; both these resistances can be divided into tolerance and avoidance mechanisms (Figure 4.17). Tolerance occurs when the cytoplasm itself is resistant; avoidance involves alternative mechanisms which prevent the cytoplasm from being subjected to the damaging temperatures.

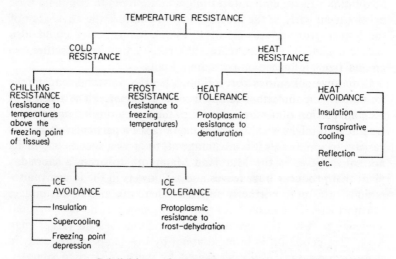

FIGURE 4.17. Subdivisions of temperature resistance, based on Levitt (1958) and Larcher (1973b).

4.4.1 *Possible mechanisms of temperature resistance*

There is no single mechanism that confers temperature resistance on plants. Many plants show maximum resistances to heat, cold and drought during winter and thus a consideration of the possible mechanism of cold resistance has a general interest. Larcher (1971) divides this into three major stages, each successive stage conferring a degree of resistance and being a prerequisite for the next.

The first stage consists of a retardation of growth either endogenously or through the action of some environmental factor such as temperature or photoperiod. The application of biochemical growth retardants has the same effect. When this phase is complete there is an inhibition of metabolism and entry into a dormant state.

The second stage is initiated by low temperatures (between $-5°C$ and $+5°C$) and is characterized by the reorganization and synthesis of RNA and enzymatic protein (isoenzymes) with associated changes in protoplasmic structure. Increases in heat tolerance in summer that occur in some species are possibly due to similar responses to high temperatures ($25-35°C$). Low temperatures induce the production of greater relative amounts of soluble carbohydrates and polyalcohols in the cell sap. Tissues are rendered more liable to supercool and are able to resist the sudden desiccation of cell contents by extracellular ice formation.

The third and final phase of complete tolerance is initiated by protracted freezing or by temperatures in the region of $-10°C$ to $-30°C$ or lower. The structure of bound water is stabilized and severe desiccation through freezing can be tolerated. Plant tissues in this state can be immersed in liquid nitrogen without harm.

The scheme outlined is not necessarily definitive, but it makes a nice summary of existing knowledge. The separation into three phases explains the dissimilar responses of different species and the fact that there are sometimes similarities between the resistance to cold and other resistances, such as those to heat and drought.

4.4.2 Heat resistance

Plant organs usually suffer heat damage at temperatures between $40°C$ and $55°C$ and at first sight such temperatures appear unlikely to occur in nature. However, a consideration of the energy balance of leaves leads to the conclusion that the thermal death of leaf tissue is not an unreal danger, even in a temperate climate, and if transpiration is low, then high leaf temperatures can rapidly be attained (Figure 4.18). In cold climates this may be an advantage.

Heat damage may also be caused by fire, for while parts of the plant may be totally consumed lower temperatures are produced in deeper tissues of massive organs and in the below-ground parts of plants. Many tropical ecosystems (e.g. savanna) and temperate vegetation types (e.g. heath and chaparral) are maintained by fire, and thus the study of heat resistance has both an ecological and an applied relevance.

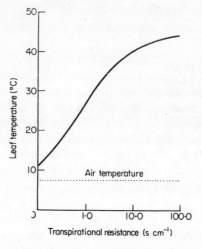
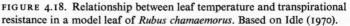

FIGURE 4.18. Relationship between leaf temperature and transpirational resistance in a model leaf of *Rubus chamaemorus*. Based on Idle (1970).

Plants may avoid heat damage by a variety of means (Figure 4.17). Leaves may be oriented parallel to the incident radiation or may have a high reflectivity and thus reduce the amount of energy absorbed. The absorbed heat may be efficiently dissipated by high transpiration rates or, in small leaves, by efficient convection and so maintain the leaves at a sublethal temperature. Buds and cambial tissue may be insulated within other tissues and thus survive high external temperatures, such as those which occur in fires. Plants also avoid the effects of high temperature by being confined to cool habitats, e.g. water, shade.

Paradoxically, the heat tolerance of species from temperate and arctic regions is often greatest in winter and appears to be associated with the degree of winter dormancy (Table 4.3). There may be adaptive increases in heat resistance during the summer but these may vary with the climate and geographical location of the species. Shoots of *Erica tetralix* from central Germany show a greater increase in heat tolerance during a hot summer than in either a normal continental summer or an oceanic summer (Table 4.4). Species from climates where there is no marked winter dormancy (e.g. *Olea europea*, Table 4.3) may show a higher tolerance in summer than in winter, whilst in many algae the heat tolerance follows environmental temperatures (Larcher 1973a).

TABLE 4.3. Heat resistance in summer and winter for selected species.

	°C summer	°C winter
(a) Arctic vascular plants		
Equisetum scirpoides	52	72
Arctostaphylos uva-ursi	52	60
Ledum groenlandicum	50	56
Vaccinium vitis-idaea	48	56
(b) Northern oceanic dwarf shrubs		
Erica cinerea	47	51
E. tetralix	47	52
Calluna vulgaris	47	50
Vaccinium myrtillus	48	52
(c) Northern temperate evergreens		
Picea abies	46	47
Abies alba	46	48
Taxus baccata	50	49
(d) Mediterranean trees and shrubs		
Prunus laurocerasus	48	49
Pittosporum tobira	51	49
Quercus ilex	53	47
Olea europea	55	48

In tropical regions there is less distinction between summer and winter but the maximum value (56°C) recorded for species of both desert (Lange 1959) and rain forest (Biebl 1964) is similar to maximum values for the Mediterranean summer (d, above; see also Lange & Lange 1962, 1963).

The table uses maximum values obtained in each season and derives from (a) Riedmüller-Schölm 1974 (Alaska); (b) Bannister 1970 (Scotland); (c), (d) Pisek et al. 1968 (Austria). Note the trend of plants from northern localities or with northern affinities to show maximum heat resistances in winter whereas plants of southern origin show maximum values in summer.

Heat tolerance appears to be correlated mainly with the stage of growth of the plant and young, actively growing, tissues are less tolerant than older, dormant, tissues (Lange & Schwemmle 1960), Lange 1961). An environmental parameter such as daylength, which is associated with the growth and development of most plants, may be also strongly correlated with season changes in heat tolerance (Bannister 1970) and plants which are experimentally subjected to constant temperature and continuous illumination show some dampening of the annual course of tolerance (Schwarz 1970). The water content of tissues also varies over an annual cycle (Figure

TABLE 4.4 Seasonal variation of the heat resistance of *Erica tetralix* from Germany and Scotland. (Derived from Lange 1961 and Bannister 1970).

	Germany		Scotland	
	old shoots	young shoots	old shoots	young shoots
Winter maximum	50·6	—	51·9	—
Summer minimum	44·0	44·6	47·5	44·3
Summer maximum (normal summer)	49·0	47·0	49·4*	46·6
Summer maximum (hot summer)	50·6	49·0 (°C)	—	—

Heat resistance estimated as the temperature required to damage shoots exposed for 30 minutes.

*Value for June which is typically the hottest and driest month in Scotland. All other summer maxima are for August. (August value for old shoots from Scotland is 47·5°C).

Note the great resistance of older shoots and the higher resistance during a hot summer.

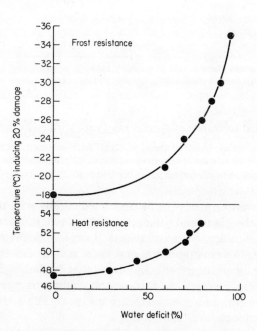

FIGURE 4.19. Effect of desiccation upon the frost and heat resistance of fronds of *Polypodium vulgare* (redrawn from Kappen 1966).

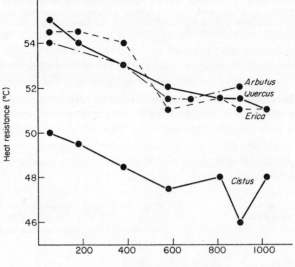

FIGURE 4.20. The heat resistance of leaves of *Cistus salviaefolius, Erica arborea, Quercus ilex* and *Arbutus unedo* in relation to their altitude of origin (after Lange & Lange 1962).

6.14) and is also correlated with changes in heat tolerance (Bannister 1970) whilst induced water deficits may considerably increase tolerance (Figure 4.19).

The heat tolerance of species may vary with environmental factors such as altitude (Figure 4.20). This may represent an adaptation to prevailing temperatures or merely reflect the slower development that occurs at higher altitudes. Within a local area, the heat tolerance may be associated with habitat, but appears to be more strongly related to the type of plant. Plants with efficient avoidance mechanisms may be less tolerant than those with limited avoidance and heat tolerance has been shown to vary with the amount of transpirative cooling that occurs (Figure 4.21). Thus, in southern Spain, the heat tolerance of mesophytes (which typically have high transpiration rates) varies between 44°C and 50°C whilst that of sclerophylls (with low transpiration rates) lies between 50°C and 55°C (Lange & Lange 1963).

There are also some geographical trends in heat tolerance as the highest values in summer are found in warmer and more southern localities (Table 4.3). The heat tolerance of species which exhibit avoidance may show little correlation with geographical location, as, for example, in species of deep shade or from underwater which may be susceptible to temperatures as low as 40°C (Sapper 1935).

The high heat tolerance of many plants in winter would not appear to have much ecological relevance and suggests a similar mechanism to that involved in the development of frost tolerance. However, the adaptive increases in tolerance that occur in summer or in response to artificially imposed temperatures (Lange 1962) are more eco-logically relevant and suggest that different changes in the structure and chemistry of cells must occur although these may be similar to changes induced by summer drought.

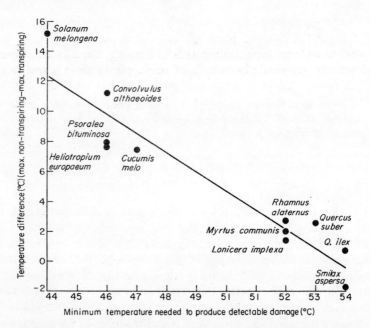

FIGURE 4.21. Relationship between the degree of transpirative cooling and heat resistance of Mediterranean plants (based on data of Lange & Lange 1963).

4.4.3   Resistance to cold

Frost resistance has been more extensively examined than heat resistance (Alden & Hermann 1971), possibly because of the extensive damage caused to plants of economic importance during spells of cold weather. Death in winter is often attributed solely to frost, but it is likely that desiccation also plays a part. This interaction is recognized in the term *Frosttrockniss* (frost-drying) used by German workers (e.g. Michael 1967). Even when frost alone is responsible for death, the conversion of tissue water to ice causes desiccation of the protoplasts as well as mechanical damage.

Avoidance mechanisms are a component of frost resistance (Figure 4.17). Many plants avoid the cold season of the year by developing from seed or some perennating structure which is buried beneath the ground and protected from frost. Some physiological changes that allow plants to endure frosts are avoidance mechanisms because they prevent the formation of ice within the tissues. These include the production of osmotically active substances (e.g. sugars) that lower the freezing point of the cell sap and the capacity of tissues to undergo supercooling without freezing. The frost resistance of actively growing plants and of many plants from warm (e.g.

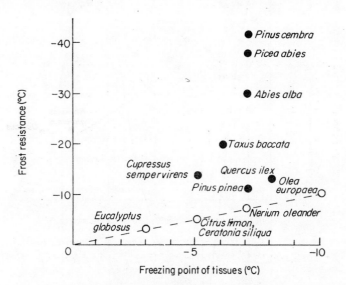

FIGURE 4.22. Relationship between frost resistance and freezing point in a variety of species (data of Larcher 1973).
● ice-tolerant species, ○ ice-sensitive species.
The dashed isometric line indicates where frost resistance and freezing point are identical.

Mediterranean) climates is brought about by this type of physiological avoidance which does not operate at temperatures much below − 10°C (Figure 4.22).

Some algae, fungi and higher plants from the humid tropics may be injured by temperatures above freezing. Most plants will tolerate some frost but are susceptible to ice formation within their tissues whilst some, particularly in winter, are resistant to formation of ice.

The annual course of frost resistance generally shows a winter maximum and a summer minimum although there may be considerable differences in the seasonal amplitude of species growing on the same site (Figure 4.23). It is often correlated strongly with both photoperiod and temperature and frost resistance can be induced experimentally by subjecting plants to short days and low tem-

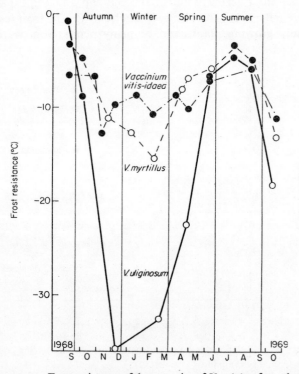

FIGURE 4.23. Frost resistance of three species of *Vaccinium* from the same upland site. Data of Polwart (1970).
● leafy shoots, ○ leafless shoots.

peratures (Polwart 1970). The annual amplitude of frost resistance is considerably dampened if plants are kept under constant temperature and illumination (Schwarz 1970).

The tolerance of cold, like the tolerance of heat, appears to be correlated with the stage of growth of the plant. Dormant tissues are normally the most resistant whereas young, rapidly growing, tissues are very susceptible to frost; thus shoots which may survive temperatures of less than $-30°C$ in winter may be susceptible to slight frosts during the growing season (Figure 4.23). The presence of water deficits usually increases frost resistance (Figure 4.19).

TABLE 4.5. Cold resistance of leaves of groups of plants from various climatic zones (based on Larcher 1973a).

	range (°C)
Tropical herbs and trees	$+5 - -2$
Subtropical shrubs and succulents	$-8 - -12$
Temperate shrubs from oceanic sites	$-6 - -15$
Deciduous trees and shrubs of wider distribution	$-25 - -40$[*]
Temperate herbs	$-10 - -20$
Water plants	$c. -10$
Boreal conifers	< -40
Alpine dwarf shrubs	$-20 - -70$
Arctic-alpine herbs	$-30 - -196$[*]
Arctic vascular plants	$0 - -80$[1]

*Resistance of buds.
[1]Riedmüller-Schölm (1974).

In most cases (Table 4.5) the degree of frost resistance developed in the winter is far in excess of the lowest environmental temperatures and it is the rate of 'hardening' in autumn and 'dehardening' in spring which determines the susceptibility of species to frost. In Norway, the native spruce (*Picea abies*) is more resistant to low temperatures than the introduced douglas fir (*Pseudotsuga menziesii*) and sitka spruce (*Picea sitchensis*) and the Norway spruce is found to commence growth later and cease earlier than the other two species (Oksbjerg 1966). The maximum cold tolerance developed in populations from different locations may be very much the same (Smithberg & Weiser 1968, Flint 1972), but the timing of phenological events causes differences in susceptibility. In general, frost resistance is more closely related to phenology than to the time of year, particularly when populations from different environments are being compared

(Larcher and Mair 1968). If a plant is prevented from maturing during the growing season then its susceptibility to frost is increased. *Vaccinium myrtillus* from a lowland habitat is more resistant to winter frost than material from above the treeline—this may be associated with the shorter growing season at higher altitude and may also explain the confinement of upland communities of *Vaccinium* to areas of snow lie (Figure 4.24).

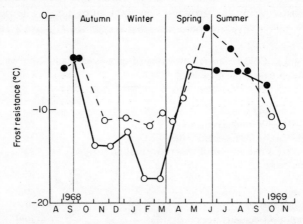

FIGURE 4.24. Frost resistance in *Vaccinium myrtillus* from an upland site (655m a.s.l, ---) and a lowland site (105m a.s.l., ——). Data of Polwart (1970).
● leafy stems, ○ leafless stems. Vertical lines indicate the position of equinoxes and solstices.

The frost resistance of species may be related to their geographical distribution, even when the species are growing in a similar habitat. Thus *Vaccinium uliginosum*, which has a more northerly distribution than *V. myrtillus* and *V. vitis-idaea*, has the greatest frost resistance (Figure 4.23). On a broader geographical basis there is a good correlation of frost resistance with origin. Some plants from the tropics are susceptible to temperatures above freezing whilst many plants from regions with cold winters may be able to tolerate temperatures of less than −40°C. Frost resistance is also correlated with lifeform and species with exposed buds are generally more resistant than those with protected buds (Table 4.6). Within a group of a particular life form, species characteristic of areas with milder climates usually have a lower resistance than those which can be found in areas with severe winters, even though they are growing together in the same environment.

Species seem to be well adapted to survive the low temperatures that may occur in their particular habitat and climatic zone. The influence of temperature and photoperiod on the development of

TABLE 4.6. Maximum frost resistance of various plant organs investigated in Central Europe (Till 1956).

	Temperature range (°C)
Underground organs (Roots, rhizomes and buds)	−6·0−−13·5
Above-ground organs	
Evergreen leaves	
3–5 cm above the litter layer	−11·5−−14·5
5–10 cm above the litter layer	−11·5−−18·0
10–20 cm above the litter layer	−13·0−−20·0
Buds of herbaceous plants	
under litter (1–2 cm above the ground)	−7·0−−11·5
at litter level (2–4 cm above ground)	−12·5−−15·5
above the litter (5–20 cm above ground)	−15·0−−19·5
Trees and shrubs	
E. tetralix (leaves and buds)	−19·5−−20·0
Taxus baccata (needles)	−35·0
Buds of deciduous trees	−21·0−−40·0

resistance helps to ensure this. It is only exceptionally that natural populations are damaged by frost although the dying back of some species, such as ferns, in autumn is associated with the first frosts. When frost damage occurs in the field it is possible to relate the susceptibility of species to resistances determined in the laboratory. There is a remarkably good agreement in the case of ferns (Bannister 1973). When plants are transplanted from their normal environment they then become susceptible to frost damage if they cannot develop adequate resistance in their new environments. Plants which are intolerant of ice formation within their tissues will not be able to survive temperatures that would allow this, whereas the timing of phenological events in a new environment may render the transplanted species susceptible to frost early and late in the growing season.

5: The Soil Environment

Previous chapters have considered the aerial environment and its effect upon plant responses to light and temperature. However, many plant functions are influenced by the environment of the soil: the uptake of minerals is almost wholly concerned with the soil, whilst plant water relations are influenced by both the soil and the aerial environment. Consequently, this is an appropriate stage at which to consider the soil environment in greater detail so as to provide the essential background for the following chapters on water relations and mineral nutrition.

5.1 PHYSICAL ASPECTS

5.1.1 *The thermal properties of soils*

The energy budget of a soil surface can be analyzed in exactly the same way as that of any other surface (2.1.3). The soil receives both long- and shortwave radiation which is variously absorbed, reflected and reradiated as well as being used in the evaporation of water and lost or gained by convection.

However, although the ecologist has a considerable interest in the energy relations of the soil surface, he also is concerned with the transfer of heat and the temperature of the bulk of the soil. The thermal properties of soil can be described by two physical parameters, its specific heat (or heat capacity) and its thermal conductivity.

The specific heat is the amount of energy needed to raise a given amount of soil by 1°C and is usually measured in J g^{-1} °C^{-1} in SI units (formerly as cal g^{-1} °C^{-1}). However, the expression of the specific heat of soils or soil constituents on a gravimetric basis can be misleading. For example, the specific heat of air is about 1·0 J g^{-1} °C^{-1} and that of water about 4·2 J g^{-1} °C^{-1}, but air has a very low density and its volumetric specific heat is about 1·2 kJ m^{-3} °C^{-1} as compared with 4·2 × 10^3 kJ m^{-3} °C^{-1} for water. The volumetric specific heat for other soil components varies around $2\cdot0 \times 10^3$ kJ m^{-3} °C^{-1} (Table 5.1). Soils of low bulk densities, such as peats, have a high gravimetric specific heats but when expressed volumetrically their specific heats are comparable to those of other soils. The specific heat of a soil is largely determined by its water content (Figure 5.1). If a soil with 50 per cent pore space is saturated with water then its volumetric specific heat will be about three times as great as when it is completely dry. If the same saturated soil is frozen then it will have a lower specific heat because the specific heat of ice is less than half that of water (Table 5.1). In cool temperate climates soils with a high retention of water (e.g. clays) are often termed 'cold' soils as they are slow to heat up and thus make the soils unsuitable for rapid plant growth early in the growing season. In hotter climates a high water content allows evaporative cooling and prevents such soils from becoming too hot.

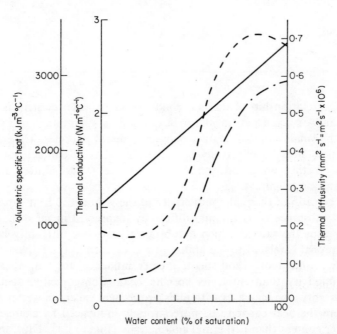

FIGURE 5.1. Relationships between specific heat, thermal conductivity and thermal diffusivity in a hypothetical soil (derived from Kohnke 1968). ———— specific heat, —.—.— thermal conductivity, — — — thermal diffusivity.

The other major parameter in soil thermal relations is *thermal conductivity*. This is a measure of the ability of a substance to transfer heat and is usually expressed in terms of $Js^{-1} m^{-1} °C^{-1} = W m^{-1} °C^{-1}$ (SI) ($= cal s^{-1} cm^{-1} °C^{-1}$). In soils the air has much the lowest conductivity ($0.025 W m^{-1} °C^{-1}$), the conductivity of soil minerals is about one hundred times greater than this and that of water about twenty times greater. Thus, water has a critical role to play in the conduction of heat by soils as well as in the determination of thermal capacity.

TABLE 5.1. Specific heat of some soil components (after Kohnke 1968).

	Specific heat $(J g^{-1} °C^{-1}$	Volumetric heat capacity $(J cm^{-3} °C^{-1})$
Water	4·2	4·2
Ice	2·1	1·9
Air	1·0	0·0013
Humus	1·7	2·3
Clay	0·9	2·1
Quartz	0·8	2·1

The relationship of thermal conductivity to water content is not linear. Air acts as an insulator in dry soils and conduction is only by the small areas of contact between mineral particles. The introduction of water effectively increases the areas of contact but has little further effect on conductivity as the water content nears saturation (Figure 5.1).

The ratio of thermal conductivity to the volumetric heat capacity of a substance is its thermal diffusivity (expressed in $m^2 s^{-1}$). This expresses the rate at which a substance heats up; alternatively its reciprocal is related to the ability of a substance to retain heat. The water content of a soil has a greater influence upon its thermal conductivity than upon its specific heat. Consequently, thermal diffusivity also increases with moisture content except in wetter soils where the heat capacity is more strongly influenced by changes in water content than is thermal conductivity (Figure 5.1). Thus, moist soils absorb and give up heat relatively rapidly and buffer air temperatures near the ground whilst soils of a lower thermal diffusivity allow more marked fluctuations in air temperature.

5.1.2 *Soil temperature*

The relatively low thermal diffusivity of most soils means that the transfer of heat within soils is a slow process. This leads to the establishment of marked gradients of temperature within a soil profile; during the day the maximum temperature is at the surface and the temperature generally decreases with depth whilst during the night a reversed gradient may occur in the upper layers (Figure 2.10). Also, the maximum and minimum daily temperatures occur substantially later at depth than at the surface. At only 30 cm below the surface, there may be a lag of as much as 8 hours. Hence, an afternoon maximum surface temperature will be experienced at depth near midnight whereas the dawn minimum may be recorded at the same depth only by noon the next day (Figure 5.2a).

The time of the year at which maximum and minimum soil temperatures are recorded varies with depth. The maximum temperatures at the surface are usually experienced in midsummer and are likely to occur only a matter of days later at a modest depth. At greater depth the annual range of temperatures is only a few degrees, but the maximum is displaced—at 5 m depth the annual maximum may occur in October. Thus, the deeper roots of plants may experience annual minimum temperatures during spring and maxima during autumn. Many plants do show substantial growth of roots and rhizomes in the latter part of the growing season (Ritchie 1955). However, the small amplitude of annual fluctuations in temperature at depth ensures a less extreme environment for roots than for the above-ground portions of plants.

The temperature of a soil is also influenced by the characteristics of its surface. Dark soils with a low albedo absorb a large amount of the incident energy; light soils act as reflectors. Daily maxima and nightly minima are greater in dark than in light soils, but the interaction between colour and moisture ensures that evaporation rates will be higher over a dark soil than over a light soil with a similar moisture content. This compensates for some of the difference in heat absorption and thus temperatures may not be as disparate as might be imagined.

A higher evaporation rate is likely to lead to an earlier drying of the dark soil and if this occurs temperature differences become much larger. Bare soils usually provide extreme environments, dark soils may heat up excessively, temperatures of 50°C have been recorded in coal slag heaps, and temperatures in burnt vegetation may be almost as high (Sweeney 1956). The heat load above a reflecting soil can be excessive because the plant absorbs almost equal amounts of energy from both surfaces of the leaf (Gates 1962).

Heat is not the only form of radiation that penetrates soils, light is capable of penetrating the surface although to a much lesser extent. This is most marked in coarse-grained sands where there are large gaps between the particles and the particles themselves are

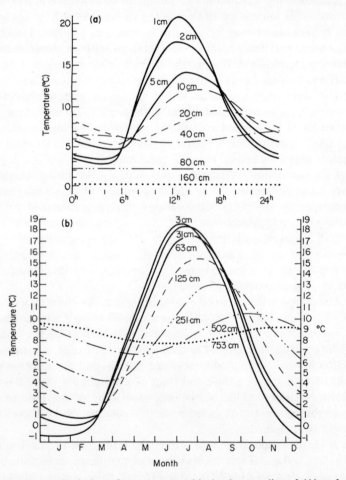

FIGURE 5.2. Variation of temperature with depth on a diurnal (A) and seasonal basis (B). Data of Schmidt 1891, Leyst 1890, after Larcher (1973).

translucent. In such sands 10 per cent of the incident light may be found at depths of 1 cm but when fine particles are included the penetration is virtually reduced to zero in the first millimetre (Baumgartner 1953). The limited amount of light penetration that does occur may be quite unimportant but it could have a significance for photosynthetic micro-organisms and for the germination of seeds buried in the superficial layers of the soil.

5.1.3 Soil structure and composition

The structure and composition of soils is determined by the various soil constituents. The mineral particles of soils are usually classified with regard to their size. Particles of diameters of less than 0·002 mm (2 μm) are considered to be clay; those with a diameter of less than 0·0002 mm (0·2 μm) are considered to be colloidal clay. Particles of diameters between 0·002 mm and 0·02 mm are classified as silt, those between 0·02 and 0·2 mm as fine sand whilst those between 0·2 and 2 mm are classified as coarse sand. The larger particles may be variously classified as gravel, stones etc.

The size of particles has important consequences. Large particles result in large pores, good aeration but poor water retention. Small particles result in fine pores, high retentivity of water but poor aeration.

Small particles have a large surface area in relation to their diameter and the *specific surface* of particles is a function of their size. This may be appreciated by a consideration of the specific surface (s) of a sphere of radius r and diameter d.

$$s = \frac{4\pi r^2}{4/3\pi r^3} = \frac{3}{r} = \frac{6}{d}$$

Thus a small spherical particle of diameter 0·0001 mm will have a specific surface of 60 000 mm^{-1} whereas a larger sphere of diameter 1 mm will have a specific surface of only 6 mm^{-1}. The internal surface of the soil is important in the short-term retention of nutrients as well as for the activity of micro-organisms. There is a vast amount of internal surface in soils; this may be illustrated by a few simple calculations. A soil of loosely packed spherical particles of 0·002 mm diameter (i.e. a specific surface of 3 000 mm^{-1}) would contain 5 000^3 = 1·25 × 10^{11} particles per cm^3. These would occupy half of the total volume (i.e. 0·5 cm^3 or 500 mm^3). The particles would have a total surface of 3 000 m^{-1} × 500 mm^3 = 1·5 m^2 per cm^3. Consequently, a hectare of soil to a rooting depth of 20 cm would have an internal surface of 1·5 × 20 × 10^8 m^2 ha^{-1} = 3 000 km^2 ha^{-1}. This is only a modest estimate as smaller particle sizes (e.g. colloidal clay) would contribute a greater amount of internal surface to the soil. The contribution of clay particles is made greater because they are not spherical but plate-like.

The relative contribution of sand, silt, and clay to the mechanical composition of the soil determines its textural classification. Soils with a good admixture of sand, silt and clay are called loams (e.g. 40 per cent sand 40 per cent silt and 20 per cent clay). When one component is in relatively greater amount the textural classification is modified; sandy loams have more than 50 per cent sand, silt loams have more than 50 per cent silt while clay loams contain more than 30 per cent clay. Soils with a preponderance of one particle size are named accordingly (e.g. silts contain more than 60 per cent silt). A convenient representation of the various textural classes is given by the U.S. Department of Agriculture soil textural triangle.

The methods of mechanical analysis usually involve sedimentation techniques; these are well known and are not described here. (See Chapman 1976).

The soil particles themselves are not the sole determinants of soil structure for the particles are held together in various ways to form larger aggregates. Both cohesive and adhesive forces serve to hold the soils together: Cohesive forces are most marked in dry soils and are due to cohesion between small particles, particularly the flat clay platelets. The cohesion is greatest when the particles are near together and decreases as they are separated. Thus as a soil becomes wetted the particles are forced apart and the cohesive forces become less marked. However, the consistency of moist soil is retained by adhesive forces, largely caused by the surface tension of water. This adhesion exerts its greatest effect in quite moist, but not wet, soils where there is sufficient contact between water and soil particles but also sufficient water/air interfaces to allow surface tension effects to develop. In drier soils the surface of contact between water and soil particles is small thus adhesion is unimportant whilst in wet soils the abundance of moisture and the lack of air does not allow water films to develop. Wet soils have almost no consistency as they are held together neither by cohesive nor by adhesive forces; such soils form slurries which resemble liquids rather than solids.

Soil becomes further structured when stable aggregations of particles are formed. The process of flocculation, where clay particles become linked to each other by the intermediacy of divalent cations, cannot account for large stable aggregates and these must be formed by further bonding or cementing. Inorganic salts such as sesqui-oxides and calcium carbonate act as bonding agents; plant root hairs and fungal mycelia bind particles together, root and microbial ex-hudates and slime capsules also help the formation and stabilization of aggregates. Aggregate formation results in the combination of the desirable characteristics of small particle sizes (large specific surface, high water retention, and high ion exchange capacity) with the advantages of large particle sizes (good drainage and aeration, lower specific heat).

5.1.4 *Soil aeration*

Soil air is of a different composition to the atmosphere. Its nitrogen content is approximately the same but the balance between carbon dioxide and oxygen varies. This is in contrast to the carbon dioxide concentration in the aerial environment which is buffered at around 300 ppm (0.03 per cent); in the soil atmosphere the carbon dioxide concentration is normally 10–100 times greater but is unlikely ever to reach 20 per cent by volume, even in anaerobic soils, because of the presence of other gases such as methane. Factors that increase the metabolic activity of soil organisms and those which restrict diffusion favour the increase of soil carbon dioxide and the depletion of its oxygen content. Thus increased soil temperature and added organic matter will tend to increase the soil carbon dioxide while the level will be decreased in soil with a high pore space and large pores.

Soil water vapour is generally abundant and at a more constant level than in the atmosphere. In soils that are moist enough to allow the uptake of water by plants, the water vapour will be near sat-uration. When reducing conditions occur in the soil, various gases such as methane, ethylene, ammonia and hydrogen sulphide (Webster 1962) may occur. These may be injurious to plants. An incubation of soils, at oxygen concentrations of < 2 per cent, for 10 days at 20°C produced concentrations of 20 ppm of ethylene (Smith & Restall 1971); such a concentration would certainly be injurious to many plants and the appearance of plants suffering injuries after flooding has been compared with the symptoms produced by ethylene poisoning. These gases appear to have a microbial origin and their production is reduced in sterilized soil.

The replacement of soil air is largely by diffusion. Mass flow is unimportant as the fluctuations of temperature and pressure that occur are unlikely to account for anything more than a small fraction (< 5 per cent) of the air in the upper areas of the soil. The diffusion of air is restricted by the sides of the pores in the soil and consequently is somewhat slower (60–80 per cent) than in free air. The rate of gaseous diffusion is directly proportional to the amount of pore space and thus the rate of diffusion is related to the water content of the soil. Soil aeration is enhanced by large particles, thus sands and soils with aggregates of particles are better aerated than silts and clays.

5.2   SOIL MOISTURE

The preceding pages have shown the importance of soil moisture in determining the storage and transfer of heat in soils. Soil moisture is also essential for plant growth and development and is the medium in which nutrients are transported and therefore an understanding of the status of water in the soil is of central importance in interpreting relationships between the plant and the environment of the soil.

5.2.1   *The water content of soils*

The water content of a given amount of soil is measured by finding the difference between its fresh (field) weight and its weight after it has been dried (usually at 105°C).

Water contents can be readily expressed either as weights or volumes and may be standardized by relating them to various soil characteristics, including fresh weight, dry weight, volume, pore space and water content at saturation. The various modes of expression give quite different values for the same soil (Table 5.2). The water content per unit pore space and per saturated water content are theoretically identical as water occupies the total pore space of a completely saturated soil. In practice, trapped air often prevents complete saturation and the saturated water content is difficult to measure as water drains from saturated soils under the influence of gravity.

However, these two measures give a scale which ranges from o per cent in completely dry soil to 100 per cent in completely saturated soil and which allows comparison between different soils. The values at saturation for other expressions will be determined by the amount of dry matter present in the soil. The water content per unit dry weight may exceed 100 per cent in organic soils (Table 5.3) and this measure is not suited to comparisons between soils. Water contents per unit fresh weight and per unit volume cannot exceed 100 per cent. The former measure has the disadvantage that fresh weight changes with water content and the latter may be less useful if the soil expands or shrinks during wetting and drying cycles.

TABLE 5.2. A comparison of various methods for the expression of soil water content.

(a) Composition of hypothetical soil

	Volume Fraction	Density (mg mm^{-3})	Weight (mg)	
air	0·25 (V_a)	0·00 (P_a)	0·00 (w_a)	
water	0·25) (V_l)	1·00 (P_l)	0·25 (w_l)	
pore space	0·50 (V_p)	—	—	
solids	0·50 (V_s)	2·50 (P_s)	1·25 (w_s)	
completely dry soil	1·00 (V_d)	1·25 (P_s)	1·25 (w_d)	
wet soil	1·00 (V_w)	1·50 (P_w)	1·50 (w_w)	

(b) Methods of expression

Basis	Equation	Values	%
per unit fresh weight	$(w_w - w_d)/w_w = w_l/w_w$	0·25/1·50	17
per unit dry weight	$(w_w - w_d)/w_d = w_l/w_d$	0·25/1·25	20
per unit fresh volume*	V_l/V_w	0·25/1·00	25
per unit pore space	V_l/V_p	0·25/0·50	50
per unit saturated weight	$(V_lP_l)/V_pP_p)$	0·25/0·50	50

*Note that the expression per unit volume may be unreliable for soils which shrink or expand during drying and wetting.

The most useful measures of the absolute amount of water in the soil are those which relate water content to the volume of the soil, as roots exploit a given volume rather than a given weight of soil. The best measures of relative amounts of water relate water contents to the available space (pore space) between the soil particles (cf. Stewart & Adams 1968). Traditional methods of measuring the amounts of soil water involve the collection of soil samples. Apart from the difficulty of taking a representative sample (most soils show a considerable point to point heterogeneity), an intensively sampled site may soon become a collection of holes. A technique that has recently found much favour is the neutron scattering technique in which a probe is placed in the soil (Stone *et al.* 1955). Neutrons are slowed down by collision with hydrogen atoms and these slowed neutrons are counted. The count is related to the moisture content of the soil although water is not the sole source of hydrogen in the soil and other atoms (e.g. iron, boron, chlorine) may interfere. The probe integrates the soil moisture content over a relatively large volume of soil and consequently minimizes errors due to small scale heterogeneity.

5.2.2   *The status of water in soils*

The measurement of water content is not as generally useful to the physiological ecologist as some measure of the availability of the water to the plant. The availability of soil moisture is a function of the forces that hold the water in the soil; the major force is due to the capillarity of the soil matrix although dissolved solutes also have a significant effect. Measures of relative water content (e.g. water content per unit pore space) indicate whether water is in short supply or abundant and allow some comparison between soils (Table 5.2). Moreover, the distribution of pore sizes in the soil will have an effect upon availability and a soil with a large number of very small pores may have less readily available water than a soil with a larger proportion of wider pores.

However, various soil 'constants' can be used to relate the amount of water in the soil to its availability. When the pore space of a soil is effectively filled with water (usually some air is trapped), the soil is said to be saturated. Water readily drains from such a soil under the influence of gravity (gravitational water) and when gravitational drainage ceases (a process that may take several days in the field) the soil is at field capacity. The water in the soil is now held by capillarity (capillary water) and the capillaries are gradually emptied as the soil dries out. Water becomes unavailable to most plants before the capillaries are completely empty and plants wilt, even in a humid atmosphere. This is the wilting point. The complete emptying of the capillaries occurs at the hygroscopic point, which marks the transition between soils which still look moist and those which appear dry. Drier soils are in equilibrium with water vapour and the remaining hygroscopic water is driven off completely by oven drying at 105°C. Any water that is still retained in soil is structurally bound to soil constituents and its removal causes irreversible changes in soil structure.

Water in the range between field capacity and wilting point is normally available to plants, but it is not equally available and is abstracted with increasing difficulty. A precise measure of the forces holding water in the soil is obtained by determining the water potential of the soil-water system. This is a thermodynamic measure which is most conveniently considered in terms of ability to do work. The effects of gravity on soil moisture are usually considered separately and thus the soil water potential can be defined as the amount of work required to transport (reversibly and isothermally) an infinitesimally small amount from a pool of pure free water to a point at the same elevation in the soil-water system. (Slatyer 1967). The units of water potential are best considered in terms of energy per unit volume as the water potential of plant tissues is usually measured in units of pressure. Water potential is thus defined as

$$\psi = \frac{(\mu_w - \mu_w^0)}{\bar{V}_w^0}$$

where μ_w is the chemical potential of the water in the system and μ_w^0 is that of pure water, while $\bar{V}_w^0$ is the partial molal volume. The chemical potential of water is designated as zero and as the water in most systems has less capacity to do work than pure free water, most water potentials are negative. Water moves down a gradient of water potential, i.e. usually from less negative to more negative potentials. The unit of water potential, as defined above, is the bar (1 bar = 10^5N m^{-2}) although the use of atmospheres of pressures still persists (1 atmosphere = 1·0133 bar). Another equivalence which has long been used in soil science is the height of an equivalent water column in cm (1 bar = 1 023 cm @ 25°C). The logarithmn of this height is known as pF (Schofield 1935). As 1 mbar is approximately equivalent to 1 cm of water, a pF is readily converted to water potential (e.g. pF 4 $\simeq$ -10^4 mbar = -10 bar).

The water potential of a soil (ψ_{soil}) is usually considered as the sum of the water potential of the water held in the soil matrix (ψ_m) and that of the water associated with a solute (ψ_s).

i.e. $\psi_{soil} = \psi_s + \psi_m$

TABLE 5.3. A comparison of the water contents of three different soils (peaty-mineral soil (a), humified peat (b), wet peat (c)).

Water potential (bar)	gravimetric (%) (water/dry wt.)			*volumetric (%) (water/volume)			relative (%) (water/saturated water)		
	a	b	c	a	b	c	a	b	c
−0·001	80	455	1 040	30	65	90	100	100	100
−0·333	48	290	520	18	41	45	61	64	50
−15·00	12	143	215	4	20	19	15	31	21
description									
gravitational	32	165	520	12	24	45	39	36	50
plant-available	36	147	305	14	21	26	46	33	29
nonavailable	12	143	215	4	20	19	15	31	21

*The volumetric water content is estimated from the density of similar soils in the field and takes no account of the expansion and shrinking of soils with changing water content.

ψ_m and ψ_s are often designated as τ, (matric potential) and π (solute or osmotic potential).

$$\text{Then } \psi_{soil} = \pi + \tau$$

When the soil water is subjected to pressure gradients it becomes necessary to include a term for pressure potential (ψ_p or P). This usually is positive, i.e. it adds to the capacity of water to do work.

$$\text{Thus } \psi_{soil} = \psi_p + \psi_m + \psi_s = P + \tau + \pi$$

The soil moisture constants such as field capacity and wilting point usually occur within a narrow range of water potential. The potential is zero in a fully saturated soil with no dissolved solutes. Field capacity is more variable, but a potential of −1/3 bar (pF 2·53) used as an approximation to the field capacity of the majority of soils. This point is often used to determine the wetter end of soil water availability as potentials greater than this are usually associated with impeded drainage and waterlogging.

The wilting point varies both with the soil and the plant species. However, soil at wilting point shows large changes in water potential for small changes in water content and thus differences in wilting point of a few bars are likely to be reflected in only small differences in the amount of water. For most practical purposes it is convenient to use a water potential of − 15 bar (pF 4·18) to estimate the wilting point. The amount of water held in a soil between potentials of − 0·33 bar and − 15 bar can be considered as available to plants.

The water potential at the hygroscopic point is around − 31 bar and that of oven dry soil around − 10⁴ bar but both these points are in a range where water is no longer available to the plant and are therefore not of particular interest to the plant ecologist.

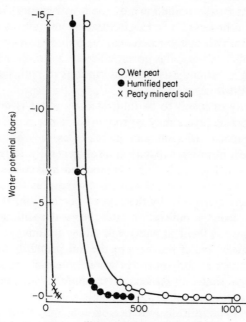

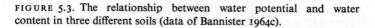
Water content per unit dry weight

FIGURE 5.3. The relationship between water potential and water content in three different soils (data of Bannister 1964c).

Some of the points made in this section may be better understood be reference to actual soil moisture curves (Figure 5.3). The gravimetric water contents of the three soils (a mineral soil, a humified organic soil and a wet peat) are widely different at any water potential, although there is a much better correspondence when all are expressed as relative water contents. (Table 5.3). The absolute amount of available water (by volume) is greatest in the more organic soils but the relative amount is less. Almost half the total water content of the mineral soil is available in contrast with a third in the organic soils. The peat contains a large amount of gravitational water whereas the well-decomposed humus soil retains a lot of water in the fine capillaries between the humus particles.

5.2.3 *The movement of water in soils*

The movement of soil water can be subdivided into two phases: the movement into soils and the movement within soils. Water moves into soils by a process known as infiltration and moves downwards in a near-saturated soil by percolation. Subsequent to infiltration the permeability of the soil is governed by its hydraulic conductivity, a term which is currently used in descriptions of both saturated and unsaturated flow.

The capacity of a soil to be infiltrated by water is not constant. Some dry organic layers may be somewhat water-repellant and the infiltration capacity of such soils increases as they become wetted. However, the infiltration capacity of a soil generally decreases during a period of rain. The soil becomes compacted, soil colloids swell and the entry of water is restricted causing the surplus to run off. Infiltration is also determined by the rate of percolation. If the percolation rate exceeds the infiltration rate, then water will move into the soil; otherwise the incident water will either accumulate on the surface or run away. Water moves into the soil primarily under the influence of gravity as differences in water potential are small in the saturated layer that is produced after rainfall. If a puddle is produced, then water may move into the soil under positive pressure.

The rates of infiltration and percolation are of the same order and rapid rates may exceed 250 mm hr^{-1} whereas slow rates may be less than 5 mm hr^{-1}. These extremes may be found in the same soil, for example, a clay with a prismatic structure would initially allow rapid infiltration between the aggregates but would swell as it became wetted and infiltration would become drastically reduced. Infiltration is usually better under vegetation than on bare areas, but is such a variable characteristic that few generalizations can be made.

Water moves through the soil in response both to gravity and differences in water potential. These may be combined to form a total soil water potential (Ψ) (Slayter 1967)

$$\Psi = \psi_{soil} + \psi_z \text{ where } \psi_z \text{ is gravitational potential.}$$

In saturated soil, the amount of water moving through a uniform cross-section of soil in unit time (v) is proportional to the difference in hydraulic head (D'Arcy's Law). In terms of total soil water potential this gives

$$v = -K\frac{\Delta\Psi}{z}$$

where K is the hydraulic conductivity and z is the depth in the soil. This equation can be modified for use in unsaturated soil, where the gravitational potential becomes less important and movement is more in response to differences in water potential. The hydraulic conductivity of unsaturated soils is less than in saturated soils and the degree of difference is related to the mechanical composition of the soil. Coarse soils have a high conductivity when wet but water surfaces become discontinuous as they dry and the large pores become emptied, consequently the conductivity decreases. Soils made up of small particles will have a low conductivity, but the small pores which account for this also ensure a prolonged retention of continuous water surfaces, and thus smaller decreases in conductivity, as the soil dries.

The movement of water into plant roots is a function of the difference in water potential between the root and the soil water. In wet soils a small difference is sufficient to allow uptake but in drier conditions the poor hydraulic conductivity of many soils may result in a desiccated zone forming around the root, (Tinklin & Weatherley 1966). The rate of flow can then be maintained only if there is a considerable decrease in the water potential within the root. It is difficult to assess the importance of lateral movement of water in such conditions and it is possible that the rate of growth of roots in dry soils may outstrip the rate of water movement.

5.3 CHEMICAL ASPECTS

Soil is formed over a period of time as a result of interactions between inorganic parent material, climate and vegetation. Parent materials are weathered by climate and their breakdown is assisted physically by plant roots and chemically by exudates and the products of the decomposition of plant residues. Plants also add an organic component to the soil. The parent materials and the plant residues are the major sources of minerals in the soil although there are also additions from minerals dissolved in rainwater and moving groundwater (Figure 5.4). The availability of minerals to the plant will be determined by both the chemical and physical nature of the soils and also by its biological activity.

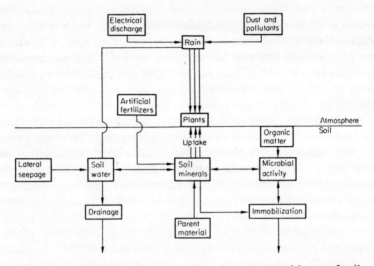

FIGURE 5.4. Simplified diagram to show the sources and losses of soil minerals taken up by plants.

5.3.1 *Soil minerals and soil formation*

The weathered parent rocks form the prime source of minerals for the soil. These may be broken down *in situ* (forming residual soils) or the derived material may be transported by various agencies such as water, wind, ice (glaciers) or gravity.

The minerals in the soil are rarely simple and the majority are complex aluminosilicates, containing various metals, that can be broken down with varying degrees of ease. One of the ultimate products is quartz (silicon dioxide, SiO_2) which is very common in soils from temperate regions. Other common minerals are felspars which break down slowly and provide sources of potassium, sodium, calcium and aluminium; micas which break down more readily and provide potassium, magnesium, iron and aluminium; and pyroxenes which also break down fairly readily and yield calcium, magnesium and iron. A variety of other minerals also occur including compounds of iron such as ilmenite, haematite and magnetite; apatite (calcium phosphate), which is a source of phosphorus; and more exotic minerals such as zircon and garnet.

The weathering of many of these compounds produces clay minerals. These are colloidal and have a plate-like structure with two principal types—kaolinite and montmorillonite. These are oxides of silicon and aluminium but other elements can be contained within their crystal lattices. The clays usually have a residual negative charge and this allows them to hold cations. Divalent cations (e.g. Ca^{++}) may bind plates together. The chemistry of these clays is dealt with more fully in textbooks on soil science (e.g. Russell 1974).

Both the parent rocks and weathered material are soon colonized by organisms. Lichens are typical early colonizers and aid weathering by retaining water, by chemical exudation and by the solution of respiratory carbon dioxide. Their death and decay add organic matter to the developing soil. The organic increment becomes larger as colonization proceeds and soil formation eventually becomes more under the influence of biological rather than physical factors. Plant roots break up the rock, exudates and the products of organic decay aid the dissolution of minerals, whilst the addition of organic matter alters the physico-chemical nature of the soil. The direction of soil development is strongly influenced by climate but also by local topography and geology and by the length of time that the soil has been developing. These considerations are the basis of a broad classification into zonal soils, where gross climate has a major influence; intrazonal soils where local factors predominate; and azonal soils which have had insufficient time for development.

Soils and soil formation have been a prime interest of many ecologists, because soil, like ecology itself, is the expression of many interacting factors. Consequently, some ecological texts (e.g. Etherington 1975) deal with soil in detail; however, a detailed consideration is beyond the scope of this text and what follows is the barest of outlines.

The major climatic factors influencing the development of soils are precipitation, evaporation and temperature. An excess of precipitation over evaporation results in a net downward movement through the soil and the loss and redeposition of materials through leaching. A converse ratio results in the upward movement of salts and their concentration in the superficial layers of the soil. Temperature and aeration influence the rate of decomposition of organic matter so that it accumulates in cool and wet climates and disappears rapidly in warm and dry ones. The vertical movement of water and the leaching and deposition of materials that occur in soils result in the formation of distinct horizontal layers, or horizons, in many soils. The horizons are labelled alphabetically from the soil surface: **A** horizons are often leached of some materials although they may be enriched in organic matter; **B** horizons are often zones of redeposition; whereas **C** horizons are usually unaffected by the processes of soil formation and may or may not be the parent material. The **R** horizon is the underlying rock. Other horizons may include organic (**O**) horizons which are usually found above the **A** horizon and gleyed, **G**, horizons where waterlogging induces the formation of reducing conditions. Subdivisions within horizons are usually denoted by suffixes (e.g. A_1, B_2) and suffixes may also be used instead of **O** and **G** to denote organic and gleyed horizons (e.g. A_o, C_g).

In north temperate Europe the climate tends to produce leached soils such as the podsol and leached brown earths (Figure 5.5) as a result of the movement of water downward through the soil profile. A similar downward movement of water occurs in the humid tropics, but while tropical podsols may occur on well-drained materials such as siliceous sand, most tropical soils are characterized by the mobilization of silica and silicates leaving clay and iron and aluminium oxides; in contrast to podsols where the acid conditions mobilize clay and organic matter and silica remains. Other soils show less downward movement of water. Tundra soils are wet and organic and well mixed by frost heaving; drainage is impeded by permafrost and organic decomposition is slow because of the cold and wetness. Steppe soils are also frozen in winter but the hot arid summers cause an upward movement of water so that such soils contain free calcium carbonate and are black with well-humified organic matter. These are the chernozems or black soils of what are now the wheat belts of Russia and North America. In more arid regions they give way to soils with more free salts and less organic matter.

The most prevalent local, or intrazonal, soil types are caused by lack of drainage and specific parent rocks. Waterlogging produces gleyed soils or horizons where reducing conditions predominate. Organic matter accumulates, particularly in acid soils, although it

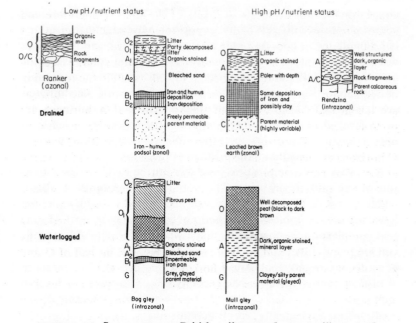

FIGURE 5.5. Some common British soil types chosen to illustrate the effects of climatic and local factors on soil development.

may be well decomposed in soils of high pH or nutrient status (Figure 5.5). Special soils are formed over calcareous rocks. In temperate regions a well-decomposed organic soil rich in calcium carbonate (a rendzina) is formed directly over the parent rock (Figure 5.5). In more arid climates this type grades into the terra rossa and terra fusca of the Mediterranean regions; the characteristic red and brown colorations being produced by enrichment with iron oxides. Specialized soils are developed in a variety of other conditions, for example on saline substrates.

Azonal soils show almost no profile development and are formed on recently deposited materials such as alluvium and wind-blown sand or as an organic layer resting directly upon relatively unweathered rock (a ranker, Figure 5.5).

5.3.2 Soil organic matter

Soil organic matter is mostly derived from the vegetation and is broken down by bacteria and fungi, assisted by the activities of larger animals such as earthworms. The micro-organisms may release inorganic nutrients from the organic material (mineralization) but equally may make nutrients unavailable by incorporating them in their own structure. In well-aerated soil of mild reaction a well-decomposed mull humus is formed whereas a more fibrous mor humus is developed in acid conditions and partially decomposed layers of peat are built up in waterlogged sites. Soil organisms, such as earthworms, incorporate and mix the organic matter in less acid soils: such activity is partly responsible for the lack of marked horizons in soils such as brown earths (Figure 5.5).

The ultimate decomposition products can be divided into humin and humic and fulvic acids. The humins are insoluble in alkali, humic acids are soluble in alkali and precipitated by acid whilst the fulvic acids are soluble in alkali and not precipitated by acid. Humic acid complexes are spherical in shape whereas those of fulvic acids are long and filamentous. Both are colloidal and, like the clay minerals, are negatively charged and can hold cations. The presence of organic matter increases the capacity of soil to hold cations and humus adds to the soil's capacity to hold water, although the precise relationships vary greatly with different soils.

5.3.3 Soil reaction and base status

The acidity or alkalinity of a soil is usually measured as the negative logarithm of its hydrogen ion concentration or pH. The pH of a soil is not uniform; its value in ecological studies stems from its ease of determination and its correlation with a large number of other soil characteristics.

Hydrogen ions tend to be more concentrated on or near soil particles than in the soil solution, forming a 'cloud', the diffuse double layer, around each particle. This results in the soil solution having a greater pH than the soil particles. The addition of salts concentrates the diffuse double layer and more hydrogen ions are released into solution, thus reducing the difference in pH. For this reason some workers prefer to measure the pH of soils after salts, such as potassium chloride, have been added. Other factors which influence the pH of soils are the oxidation-reduction potentials and the presence of carbon dioxide. The pH of waterlogged soils decreases as they dry out: reduced compounds such as sulphides are oxidized to sulphates with a consequent reduction in pH. The influence of carbon dioxide is most strongly felt in soils with a pH greater than pH7. In calcareous soils the maximum pH is limited by the chemical equilibrium between carbon dioxide, carbonate and bicarbonate ions and cannot exceed pH8·5. Soils of higher pH, alkali soils, usually have free sodium carbonate but their pH rarely rises above pH9·5.

Low pH values (< pH4) are usually associated with free acids. Hydrated aluminium ions usually buffer soils against excess acidity so that pH values in mineral soils rarely fall much below pH4. The buffering capacity of the hydrated aluminium ions is due to their ability to dissociate or gain H^+ ions as the pH rises and falls. Different soils have different buffering capacities (Figure 5·6) and this may be related to their composition—clays would be expected to have a higher buffering capacity than sands.

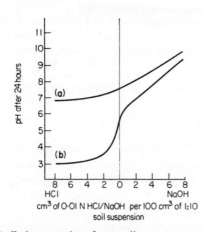

FIGURE 5.6. Buffering capacity of two soils.
(a) Supporting *Festuca vaginata* (b) Supporting *Corynephorus canescens* (based on Rychnovská 1963).
Corynephorus is confined to soils of low buffering capacity (see 7.1.3).

The pH of a soil determines the availability of ions (7.1.2). Soils of low pH are infertile as the dominant cation is H^+ and there is little room for other cations. Elements such as iron, aluminium and manganese are available under acid conditions whilst other, such as calcium, magnesium and molybdenum show an increasing availability with increasing pH (Figure 5.7).

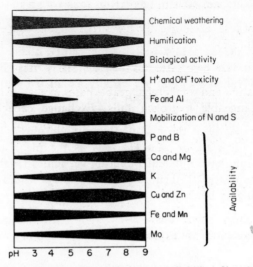

FIGURE 5.7. Variation in soil properties and availability of ions in relation to pH (after Schroeder 1969 and Larcher 1973a).

The breadth of the band is related to the degree of activity of availability (i.e. broad bands indicate high availability).

5.3.4 *Ion exchange*

Most of the readily available ions in the soil are attached to soil particles although some are in solution. Clay and organic particles are negatively charged and thus can bind cations. In clays the negative charges result from electrical imbalance caused by the replacement of metallic ions within the crystal lattice and from the dissociation of hydroxyl groups at the edges of particles. The negative charges of humus particles result from the dissociation of hydroxyl groups at high pH ($>$ pH6) and from the dissociation of carboxyl groups at lower pH. Soils have also an ability to hold anions which appears to be associated with certain iron compounds and the presence of some basic groups in organic complexes.

The ability to hold cations is usually expressed in terms of milli-equivalents per unit soil and can be assessed by titrating the soil with dilute acids and bases (e.g. Brown 1943). Acids displace the adsorbed cations whilst bases replace the hydrogen. Thus exchange capacity is made up of total exchangeable hydrogen (TEH) and total exchangeable bases (TEB).

i.e. exchange capacity = TEB + TEH

The ratio of total exchangeable bases to exchange capacity is known as base saturation

i.e. base saturation = TEB/(TEB + TEH).

Base saturation and total exchangeable bases increase with pH whilst total exchangeable hydrogen falls. Exchange capacity is often associated with the organic content of a soil and, as acid soils are often organic, may decrease with increasing pH. The relationships between TEH, TEB and pH become complicated when the organic contents of soils are associated with changes in pH. An organic soil of low pH may have more exchangeable bases than an inorganic soil of higher pH and, as organic content is inversely related to bulk density (Jeffrey 1970), such discrepancies are accentuated when exchange capacities are expressed in terms of soil dry weight.

5.3.5   The supply of mineral nutrients to plants

Mineral nutrients may be found in the soil as ions in the soil solution, as ions adsorbed on soil colloids and as relatively unavailable minerals or complexes. The various states are not independent, a dynamic equilibrium exists between them. The ions in the soil solution are readily absorbed, those on the soil colloids interchange readily with the soil solution and represent a short-term buffer, whereas long-term supplementation is provided by slow release from soil minerals and by the mineralization of organic complexes by microbial action. Nutrient elements do not necessarily exist in all these phases, a point which is well illustrated by a consideration of three major nutrient elements—nitrogen, phosphorus and potassium.

Most of the nitrogen in soils is relatively unavailable as organic nitrogen of uncertain chemistry. About half of this can be hydrolyzed to amino acids although proteins as such have never been isolated from soil. As most of the nitrogen in soils is in organic forms, factors such as low temperature and high moisture contents which favour the accumulation of organic matter result in increased nitrogen content of soils, even though this nitrogen is largely unavailable. Most plants absorb nitrogen as nitrate from the soil solution although some plants of acid soils may utilize ammonium ions (cf. Figure 7.6). The concentration of nitrate in the soil solution is low, no more than a few parts per million, in contrast to the total soil nitrogen which is usually several hundred times higher. The nitrate in the soil solution is readily used up and is gradually replaced by the mineralization of organic nitrogen through microbial activity. The primary product is ammonia which is successively oxidized to nitrite and nitrate. Not all the inorganic nitrogen produced in this way is available to plants as much may be re-absorbed by micro-organisms. The production of surplus amounts occurs only when the ratio of carbon to nitrogen in the organic material is relatively low ($< 33:1$). Nitrate may be immobilized by the reverse process to mineralization and is also lost by denitrification in anaerobic soils where some bacteria utilize nitrate as a hydrogen acceptor and produce nitrous oxygen or nitrogen. Lack of oxygen also prevents the mineralization proceeding beyond the production of ammonia; this may be lost by leaching as ammonium ions are readily displaced by others such as manganous and ferrous ions. The source of soil nitrogen is thus entirely organic and even the major input from inorganic sources, through nitrogen fixation, ultimately depends upon the mineralization of dead organic remains.

Much of the soil phosphorus is, like nitrogen, unavailable in an organic form but this element is also immobilized as inorganic compounds such as apatite and other phosphates of calcium in soils of moderately high pH and as iron and aluminium phosphates in acid soils. These inorganic phosphates are a major source of soil phosphorus and there is no comparable source of inorganic nitrogen. Phosphorus, unlike nitrogen, is also available as exchangeable ions adsorbed on soil colloids; in this form it can be readily exchanged with the soil solution where it is present in very low concentrations (usually less than 0.1 ppm). As this phosphorus in solution is rapidly used by plants, the overall availability of the element is limited by the rate of transfer from the labile pool to the solution. Severe depletion of the phosphorus in solution and labile form is likely to be ameliorated by dissolution from more unavailable forms whereas the addition of excess amounts results in an increase in immobilized phosphorus as well as losses by leaching. These interrelationships may be illustrated by the use of a well analogy (Figure 5.8).

Potassium is found in relatively large amounts in most soils. Most of this is immobilized in various aluminosilicates which ultimately derive from feldspars and micas of the parent rocks. The exchangeable and solution fractions are relatively small and the amount of potassium adsorbed on the exchange system is generally exceeded by both magnesium and calcium.

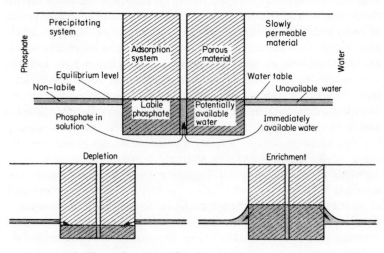

FIGURE 5.8. The well analogy for the availability of phosphate (after Sutton & Gunary 1969).

The water in the well represents the nutrient in solution and that in the porous material the exchangeable phosphate. The rate of replenishment of the water in the well (or the phosphate in solution) is a function of the head of water in the porous material (or the amount of labile phosphate). Severe depletion of the reserves of the well results in movement from the less permeable surrounds while enrichment causes the reverse movement. This is comparable with the release and immobilization of less available supplies of phosphorus. The analogy is readily modified for nutrients other than phosphorus.

Nitrogen, phosphorus and potassium are differently distributed between the various phases that can exist in the soil. Nitrogen is found only in solution and in an organic phase whereas phosphorus is available in solution, in an exchangeable form and in immobilized organic and inorganic states. Elements such as sulphur show a similar distribution to phosphorus; others, like potassium, lack an organic phase but are found in inorganic, exchangeable and solution forms.

Adsorption on soil colloids restricts the movement of ions and renders them comparatively immobile. An immobile nutrient such as phosphate is extracted from a very narrow zone around each plant root. This minimizes competition between roots and plants with a high density of roots are more efficient at taking up such nutrients than plants with less extensive root systems. Root hairs and mycorrhizas add to the effective area of exploitation and mycorrhizas are of proven importance in phosphorus uptake (8.2.2). The form of the root system is lesser importance for more mobile nutrients such as nitrate; here the zones of depletion are broader and competition between roots soon sets in so that increases in root density have less effect upon uptake (Nye 1969). The mobility of ions in natural soils is further restricted by desiccation, as pathways for diffusion become broken and more circuitous as the soil dries out. In summary, the uptake of an ion is not only a function of the demand by the plant, it also depends upon its concentration in the soil solution, its mobility in the soil and its long-term transfer from immobilized sources.

Nutrients do not only diffuse to the roots, they are also transported by mass flow along with the transpiration stream. In leeks the mass flow component represents only a small fraction (< 13 per cent) of the total uptake (Tinker 1969), although it is theoretically possible that mass flow could be more important in situations such as dry soil where diffusion is restricted or where there is an extensive area of absorbing roots.

Once the nutrients have reached the plant root, their uptake is probably a combination of active and passive processes. Mass flow could transport nutrients as far as the endodermis but then they are forced to pass through the cytoplasm. Active transport postulates the existence of 'pumps' that employ metabolic energy to transport ions against concentration gradients. There are usually considered to be inwardly directed potassium pumps and outwardly directed sodium pumps although anion pumps have also been postulated for plant cells. If pumping occurs then other ions will move in passively to restore the electrochemical balance and their regulation is made possible by differentially permeable membranes. A further mechanism of uptake is provided by the production of ions from nonionic sources as in the production of organic acids: this enables hydrogen to exchanged for cations and bicarbonate produced by decarboxylation to be exchanged for anions.

6: Water Relations of Plants

The movement of water from the soil through plants to the atmosphere is usually considered to be a purely physical process. The flux of water across the soil-plant-atmosphere system can be related to the resistance of the system and the potential drop across it (Gradmann 1928, Van den Honert 1948). In hydrology such a gross treatment is adequate and plant surfaces serve only to cause deviations from the evaporation that is expected from an open water surface. A grass sward may evaporate less than an open water surface (Penman 1948) while a forest, with evaporation taking place at different levels within the canopy and roots showing below-ground stratification, may evaporate more (Rutter 1968). However, plants react to this inevitable loss of water in a variety of ways and their tolerance or avoidance of water stress has consequences for both physiological processes and their ecological responses.

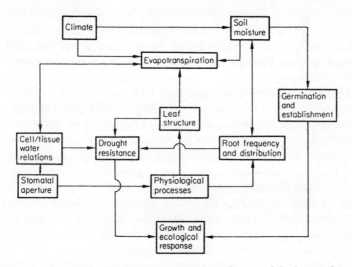

FIGURE 6.1. A diagrammatic representation of some of the interactions of plant water relations in an ecological context.

The loss of water from a plant will affect both the uptake of water and the water relations of cells and tissue. Physiological processes may be affected directly by water stress or indirectly through changes in stomatal aperture and, in turn, growth and ecological responses will be modified. Growth and developmental responses may influence leaf structure and root distribution and modify the loss of water from the plant by altering its resistance. The distribution of plants may be determined at the outset by the effects of water upon seed germination and establishment while the ability of a plant to survive in a particular habitat may be determined by its resistance to water stress and waterlogging (Figure 6.1).

The divisions made in Figure 6.1 provide the format of this chapter. The water relations of the aerial and soil environments have been considered in Chapters 2 and 5 and aspects of gaseous diffusion have been dealt with in Chapter 3.

6.1 WATER POTENTIAL AND PLANT WATER RELATIONS

Water potential has already been defined for the soil system (5.2) and can equally well describe the status of water in plant tissues, or in any other part of the soil-plant-atmosphere system. A unified terminology, based on water potential, has gained general acceptance since the recommendations of Taylor & Slatyer (1961) and has replaced the plethora of units formerly used by botanists, soil scientists and meteorologists.

Previous botanical terminologies have dated from the suction force (*Saugkraft*) of Ursprung & Blum (1916) and included the diffusion pressure deficit of Meyer (1945). These were usually expressed in atmospheres of pressure and can be readily converted into water potentials as 1 bar is equivalent to 0·987 atmosphere at S.T.P.

The water or osmotic potential of cell sap (ψ_s or π) is lower than that of pure water because of the presence of dissolved solutes. The absorbtion of water by cells causes an increase in pressure within them (turgor pressure ψ_p or P) and the water potential of the cell is the resultant of these two opposing forces

$$\text{i.e. } \psi_{\text{cell}} = \psi_s + \psi_p = \pi + P$$

In contrast to soils, the matric potential (τ) is ignored. It is difficult to assess the validity of this omission because it is not normally possible to measure turgor pressure directly (except in specialized circumstances, e.g. Arens 1939, Meidner & Edwards 1975). Estimates of turgor pressure are generally made from the difference between the water potential of a cell and that of its sap, as both these components are readily measured. The cell wall and cytoplasm may exert a matric potential and processes such as the imbibition of seeds are more readily explained in terms of matric potential.

In some instances the water potential of a tissue (which is related to the mean water potential of its constituent cells) is less than that of the cell sap. If matric potential is ignored, then turgor pressure must be negative; this implies an inward collapse of the cell walls. However such negative turgors need not exist if water is bound in a matrix, or is under tension itself, or the tension is relieved by gas bubbles coming out of solution (Milburn 1970). The concept of negative turgor is critically reviewed by Slatyer (1967).

6.2   GERMINATION AND ESTABLISHMENT OF SEEDS

6.2.1   *Germination*

The germination of seeds is generally high when water is freely available and seeds which are characteristic of plants from a range of moisture regimes usually all show optimal germination in very moist soils (e.g. Bannister 1964b). A period of soaking may enhance germination, although protracted soaking may be inhibitory. Thus, nutlets of the aquatic plant, *Polygonum hydropiper*, show maximum germination after 25 weeks of soaking at low temperature (Timson 1966) and seeds of *Erica tetralix*, a dwarf shrub often found in boggy habitats, still show a stimulation of germination after periods of soaking of up to three months (Bannister 1963). Plants of drier habitats may also benefit from a short period of soaking. *Erica cinerea*, a dwarf shrub from dry habitats, benefits from periods of soaking of up to one day although there is no germination after two weeks of soaking (Bannister 1963) whilst seeds of *Pinus strobus* show stimulation of germination by periods of soaking of up to five days (Kozlowski 1965). Stimulation of germination by soaking may be due to water leaching inhibitors from the seed, or it may merely ensure an adequate imbibition of water. Failure to germinate may be associated with the accumulation of toxic substances during anaerobiosis (e.g. ethanol, which accumulates in the roots of flooded plants —Crawford 1966) or to an increased susceptibility to fungal attack.

Presoaked seeds, or those which germinate in wet soils, are ensured of an adequate water supply for germination. However, in drier soils, the ability of a seed to make effective contact with water films in the soil is of considerable importance. Established plants have the advantage of a ramifying root system which is almost certain to make contact with water in the soil whereas seeds are small and discrete and may fail to touch a water surface when they lodge in the soil. Consequently, seeds can be shown to respond to very small differences in water tension (Huber & Merkenschlager 1951) and the area of contact between seed and substrate is of vital importance (Harper & Benton 1966). Small seeds are likely to make a more intimate contact with the substrate than larger ones, buried seeds will make a better contact than superficial ones and will also lose less water by evaporation. Small seeds, such as those of weeds, are typical of open, potentially dry situations (Salisbury 1942, Grime 1966) where their numbers aid rapid colonization but their smallness may also ensure adequate contact with the supply of soil moisture.

The response of seeds to low water tensions may be less in species from drier habitats. Thus *Jucus vaginatus* shows a marked reduction of germination at tensions of 0·6 bar in contrast to *Medicago tribuloides* (Collis-George & Sands 1959) and *Erica tetralix* is affected by tensions as low as 0·06 bar whilst the germination of *Calluna vulgaris* is unaffected (Bannister 1963). However, some seeds avoid the effects of low tension by the secretion of mucilage which effectively increases their contact with the soil (Figure 6.2).

Although germination may be reduced by low water tensions, most seeds can achieve some germination in very dry soil. Many vegetable seeds germinate in soils at wilting point (Doneen & MacGillivray 1943) and wheat seeds can germinate in equilibrium with water vapour at a potential of -32 bar and show almost 100 per cent germination at -20 bar (Owen 1952). The soil solution is concentrated in dry soils and seeds may have to abstract water against a gradient of osmotic as well as matric potential. If the seed takes up solutes or mobilizes its reserves in an osmotically active form, its own water potential will be reduced although physiological processes

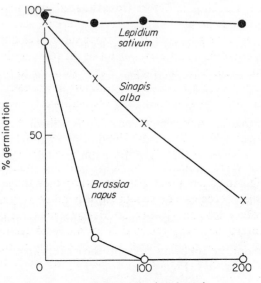

FIGURE 6.2. Effect of low water tensions on the germination of seeds with varying degrees of mucilage production. (Data of Harper & Benton 1966.)

The mucilage allows a better contact with the substrate and the ability to germinate is directly related to the amount produced. Cress (*Lepidium*) produces copious mucilage, mustard (*Sinapis*) lesser amounts and rape (*Brassica*) none.

may be inhibited both by the low water potentials and the toxicity of ions. The germination of cress, wheat and ericaceous seeds is inhibited by contact with solutions of water potentials between -2 and -4 bar (Kausch 1952, Bannister 1963) and the germination of *Iva annua*, a plant of the Nebraskan saltings, is severely inhibited by osmotic potentials of -16 bar (Ungar & Hogan 1970). In this case the seed showed similar responses to a variety of osmotica and it was concluded that the availability of water, rather than ionic toxicity, inhibited germination.

6.2.2 Seedling establishment

Once they have germinated, seedlings have to establish themselves, and it would be reasonable to expect that species are adapted to the conditions in which they are normally found. This is illustrated by the establishment of *Ranunculus* seedlings (Harper & Sagar 1953) whose response to different water tables is related to their distribution in the field. *Ranunculus bulbosus*, which is typical of ridges, shows the best establishment when the water table is low; while *R. repens*, typical furrows, responds best to high water tables and *R, acris*, which is most common on the sides of ridges, shows an intermediate response. A similar behaviour is shown by *Erica* spp.: *Erica tetralix* shows a marked reduction in establishment in drier soils, while *E. cinerea*, typical of dry habitats, shows little reduction but performs less well than *E. tetralix* on wet soils (Bannister 1964b). Species of wet habitats are sensitive to the level of the water table and more seedlings establish themselves when the water table is near the surface. Thus *Juncus effusus* shows no seedling establishment when the water table is 20 cm below the surface (Lazenby 1955) and a water table only 10 cm below prevents the establishment of seedlings of *Schoenus nigricans* (Figure 6.3).

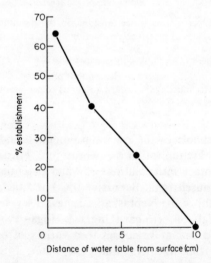

FIGURE 6.3. Effects of water table on the establishment of *Schoenus nigricans* (data of Sparling 1968).

The capability of seedlings to emerge may vary within a species or variety and is probably related to conditions during the development and maturation of the seed. Commercial seed lots which show similar, low, mortalities in standard germination tests may show radically different mortalities in the field. Those lots showing low emergence often have 'leaky' membranes and lose solutes to their surrounds: this enhances the nutrient supply in the soil and, in conjunction with the increased permeability of such seeds, renders them more susceptible to fungal attack (Figure 6.4).

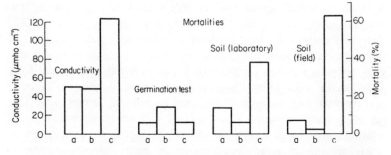

FIGURE 6.4. The conductivity of leachates of different seed lots (a,b,c) of Kelvedon Wonder peas in relation to their mortality in standard germination tests and in soil (data of Matthews 1971).

6.3   WATER LOSS AND WATER UPTAKE

6.3.1   *Transpiration and energy budget*

Transpiration is the loss of water from the leaves of plants and some of the theoretical aspects of this phenomenon have already been considered (3.1). The radiation incident on a leaf is determined by the radiation balance (S_L): transpiration (E_L) and convection (C_L) represent the principal means by which this incident energy is dissipated.

Thus $S_L = LE_L + C_L$ (where L is the latent heat of vaporization of water). If the equation is rearranged the transpiration rate can be expressed in terms of the energy budget

$$\text{i.e. } E_L = (S_L - C_L)/L$$

Consequently, transpiration will be high when convective losses are low and when there is large net gain of radiation by the leaf.

The ratio of sensible heat transfer to evaporative transfer (B) has already been given (3.1.3) as

$$B = K \frac{(T_L - T_A)}{(C_L - C_A)} \cdot \frac{R_L}{r_a}$$

where K is the psychometric constant, $(T_L - T_A)$ and $(C_L - C_A)$ are, respectively, the differences in temperature and water vapour concentration between leaf and air, R_L is the total leaf resistance to vapour transfer and r_a is the external resistance of the leaf to the transfer of water vapour and sensible heat. When the ratio is low, transpiration will be relatively high. Low ratios can be the result of large differences in vapour concentration or small gradients of temperature between leaf and air and will also occur when the total leaf resistance is low in relation to its external resistance. Thus, thin leaves with many large, open stomata would be expected to show higher transpiration rates than thick leaves with few, small, sunken stomata. Transpiration will obviously be low when stomata are closed and heat will be dissipated by convection. In such conditions large leaves, with a high external resistance, are inefficient and tend to overheat. Consequently, species with large, thin leaves may be limited to shady or moist conditions. *Impatiens parviflora*, a woodland annual, is successful in the open only on substrates such as river gravels which ensure an adequate supply of water (Rackham 1966). On the other hand, plants which conserve water by stomatal closure are adapted to avoid overheating. Succulent plants are protected by their high water content, as water has a high specific heat; many sclerophylls have a high heat resistance while the lower heat resistance of mesophytic leaves is correlated with a greater degree of transpirative cooling (Figure 4.21).

6.3.2 *Transpiration and leaf structure*

Much early eco-physiological work in plant water relations was concerned with the interrelationships of leaf structure, transpiration and plant distribution, but was unable to make any valid generalizations—even when closely related species were considered (e.g. Schratz 1932). This was possibly due to an underestimation of the importance of the ability to control transpiration and overemphasis of the actual rates of water loss, a failing that was appreciated by some contemporary workers (e.g. Maximov 1929).

Recently, the resistance to water loss has been partitioned into a number of components (3.1.2) and an appropriate equation is reproduced below.

$$R_L = \frac{1}{D}\left[\frac{\pi d_L}{8} + \frac{1}{na_s}\left(l_s + \frac{\pi d_s}{4}\right) + l_i\right]$$

where D is the diffusion coefficient for water vapour.

High transpiration rates occur when leaf resistances are low. Consequently, decreased leaf size (d_L = leaf diameter) and decreased thickness, which results in a shorter internal pathway (l_i), both lead to increased transpiration. Stomatal resistance will be lowest when there are a large number (n) of shallow (l_s) stomata with large apertures ($1/a_s \times \pi d_s/4 = 1/d_s$). The absolute amount of water lost is dependent upon leaf area, so that a large leaf with a high resistance to transpiration may still lose more water than a smaller leaf with low resistance. This has lead some workers (e.g. Heath 1969) to standardize resistances in terms of leaf area so that dimensions become s cm^{-3} rather than s cm^{-1}. However, large leaves will lose less water per unit area than small leaves.

The modifications of leaves that occur in response to different environments often result in changes in leaf resistance that do not act in the same direction. In dry habitats, plants often have smaller leaves with more stomata per unit area (Figure 6.5) and show increased leaf dissection (Lewis 1972). They therefore show a decreased resistance to water loss. This may be partially compensated for by decreased stomatal size (Figure 6.5), increases in leaf thickness and deeper stomatal pores. However, the net result is that leaves of plants from drier habitats often have potentially high transpiration rates (Figure 6.5). These adaptations appear paradoxical when viewed solely in terms of the conservation of water, as high transpiration rates imply a considerable loss. However, the reduction of the external resistance of the leaf by decreased size or increased dissection ensures that stomatal resistance becomes the overriding component when stomata are closing. Consequently, stomatal closure, when it occurs, is effective in controlling water loss. The adaptations should also be viewed from other standpoints. The decreased external resistance of the leaf results in a better coupling of leaf and air temperatures and reduces the chance of thermal damage (Lewis 1972). The increase of stomatal conductivity may be of advantage in open, drier, habitats where the maximum rate of photosynthesis is likely to be limited by the supply of carbon dioxide rather than by light intensity, and the resultant high transpiration rates will also be effective in cooling the leaves.

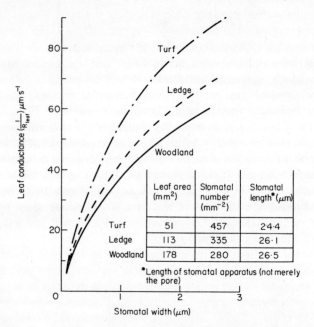

Leaf conductance $\left(\frac{1}{R_{leaf}}\right)$ μm s^{-1}

Stomatal width (μm)

	Leaf area (mm^2)	Stomatal number (mm^{-2})	Stomatal length*(μm)
Turf	51	457	24·4
Ledge	113	335	26·1
Woodland	178	280	26·5

*Length of stomatal apparatus (not merely the pore)

FIGURE 6.5. Calculated leaf conductances for leaves of *Vaccinium vitis-idaea* with the characteristics listed in the inset table. Data of Polwart (1970).

Calculations for still air at 20°C. The turf population is the most xeromorphic and the woodland population the least xeromorphic. High conductances indicate a potentially high transpiration rate.

6.3.3 *The effects of wind*

In still air, a 'shell' of humid air is formed around the transpiring leaf. This lengthens the pathway for diffusion from leaf to air and the external resistance of the leaf (r_a) is high. In wind, the shell is swept away and there is a more abrupt transition between leaf and air, with a consequently shorter pathway and decreased resistance. Transpiration is therefore potentially much higher in wind. However, stomatal closure usually occurs in response to the increased water loss and thus transpiration in the field may be little more, or even less, in windy conditions than in relatively still air. The reduction of external leaf resistance that occurs ensures that stomatal resistance becomes a greater proportion of the whole and thus stomatal closure becomes relatively more effective in controlling transpiration (Figure 3.3).

Species vary in their response to wind (Figure 6.6). Trees are normally subjected to wind and their transpiration may remain relatively constant over a wide range of wind speeds, whereas dwarf shrubs, which grow near to the ground where wind speeds are much reduced, may be sensitive to wind and react by stomatal closure which reduces their transpiration to less than that in stiller air (Figure 6.6). Stomatal closure also affects photosynthesis and wind speeds that cause complete stomatal closure may completely inhibit photosynthesis, as in *Rhododendron ferrugineum* (Figure 6.6), without completely controlling water loss as the cuticle of most plants is more permeable to water than to carbon dioxide. In contrast trees such as larch maintain open stomata and show continued photosynthesis.

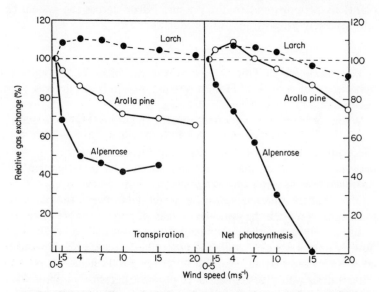

FIGURE 6.6. The effects of wind on the transpiration and photosynthesis of woody plants occurring at the alpine tree-line (after Tranquillini 1970).

Plants which are exposed to protracted periods of wind may show permanently increased rates of transpiration. This is almost certainly due to mechanical damage of the leaf surface and the stomatal apparatus. On the other hand, herbaceous plants which are grown permanently exposed to wind show increased leaf xeromorphy and a better control of water loss, although the consequence of this control is also reduced assimilation and poorer growth (Whitehead 1963).

6.3.4   *Transpiration and soil moisture*

The water potential of the atmosphere is usually very much lower than that in the soil. A soil at wilting point has a potential of -15 bar but is in equilibrium with a near saturated atmosphere; air at 80 per cent R.H and 20°C would have a potential of -300 bar. Consequently, changes in atmospheric humidity have much greater effects upon rates of water loss, and therefore on rates of uptake, than do alterations in soil moisture content. This is readily illustrated by considering the flow of water (I) through a system (i.e. plant) as proportional to the potential drop across the system (E) and inversely proportional to its resistance (R).

$$\text{i.e. } I = E/R$$

This analogue of Ohm's Law forms the basis of the catenary hypothesis of van den Honert (1948) which itself derives from Gradmann (1928).

When atmospheric potentials remain the same (Figure 6.7) alterations in soil water potential or in root/soil resistance have little effect upon the rate of loss, whilst an alteration in the resistance between leaf and air has a marked effect. The maintenance of a sustained flow of water through the plant, despite adverse conditions, implies that the difference in water potential between the plant and the soil must remain the same, or increase if resistances are increased. Thus, a doubling of the resistance between root and soil reduces the flow by only 5 per cent while the potential difference is increased by 90 per cent, and a decrease of soil water potential causing a 2 per cent reduction of water loss causes a 38 per cent decrease in plant water potential. On the other hand, if the leaf/air resistance is doubled in the same dry soil, transpiration is reduced by 50 per cent and the potential difference only increases by 1 per cent. The examples in Figure 6.7 underline the importance of stomatal movements in both the control of water loss and the maintenance of equable water potentials within the plant. Consequently, the reductions in

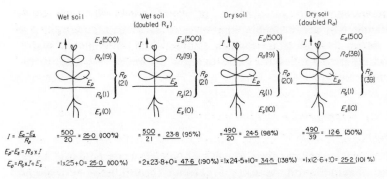

$$I = \frac{E_a - E_s}{R_o}$$ $= \frac{500}{20} = \underline{25 \cdot 0}$ (100%) $= \frac{500}{21} = \underline{23 \cdot 8}$ (95%) $= \frac{490}{20} = \underline{24 \cdot 5}$ (98%) $= \frac{490}{39} = \underline{12 \cdot 6}$ (50%)

$E_p - E_s = R_s \times I$

$E_p = R_s \times I + E_s$ $= 1 \times 25 + 0 = \underline{25 \cdot 0}$ (100%) $= 2 \times 23 \cdot 8 + 0 = \underline{47 \cdot 6}$ (190%) $= 1 \times 24 \cdot 5 + 10 = \underline{34 \cdot 5}$ (138%) $= 1 \times 12 \cdot 6 + 10 = \underline{25 \cdot 2}$ (101%)

FIGURE 6.7. A simple catenary model of the plant:soil:atmosphere system illustrating that effect of alterations in soil moisture and plant resistances upon the flow of water and water status of the plant's tissues.

$I=$ flow, R_s, R_a, R_p—total resistances of the root:soil, plant:air and soil:air systems; E_s, R_a, E_p—potentials in the soil, air and plant.

The units are arbitrary although they are of the correct order of magnitude for water potentials in the system. Potentials have been given a positive sign for ease of interpretation and calculation.

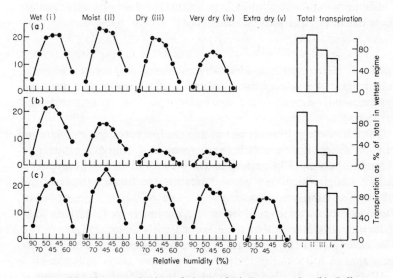

FIGURE 6.8. The transpiration of plants of (a) *Erica tetralix*, (b) *Calluna vulgaris* and (c) *Erica cinerea* in drying soil and subjected to a standard regime of atmospheric relative humidities at 20°C (Bannister 1964d).

transpiration that are observed in plants growing in drying soils (Figure 6.8) are probably due to changes in leaf resistance (i.e. stomatal closure) rather than the increased unavailability of water, and different species are likely to show dissimilar responses to the same moisture regimes (Figure 6.8). The effects of stomatal closure are considered in more detail later (6.4.3).

6.3.5   Water uptake by roots

Roots take up water from the soil when their water potential is less than that in the soil. As water is taken up, the soil becomes drier and a difference in water potential between the root and the soil must be maintained if water uptake is to continue (Kaufman 1968). The hydraulic conductivity of dry soil is very low and the depletion of water in the immediate vicinity of the roots results in the establishment of local zones of dryness. The increased resistance to uptake leads to a concomitant decrease in plant water potential (Figure 6.7). The necessary decrease is less in plants with extensive root systems which are therefore able to maintain high rates of water uptake without suffering undue depressions of plant water potential.

The uptake of water (q) may be analyzed by an expansion of the basic catenary equation (Gardner 1960)

$$q = KLA\,(\psi_s - \psi_r)$$

where K is hydraulic conductivity of the soil, L is the effective root length, A is a coefficient related to the geometry of the root system, while ψ_s and ψ_r are the water potentials of soil and root respectively.

The morphology of root systems and the frequency and distribution of roots differ considerably between species or genotypes growing in the same habitat (Figure 6.9) and are also influenced by the conditions under which they grow. There are few investigations of the effects of rooting frequency upon the water relations of plants but in *Dactylis glomerata* two subspecies with dissimilar rooting frequencies reduced soil water potential in accord with theory (Figure 6.10). The variety with the lower rooting frequency produced the largest reduction in soil water potential.

Water uptake is also influenced by the amount of root that is capable of absorption. Older parts of roots are less capable of absorption; the most important zone is only 5–10 mm behind the tip

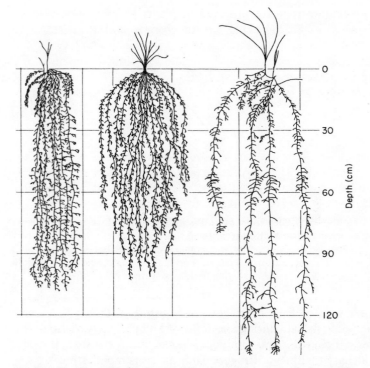

FIGURE 6.9. Root distribution in different prairie grasses. Left to right: *Calamagrostis montanensis*, *Stipa viridula* and *Calamovilfa longifolia* (Coupland & Johnson 1965).

(e.g. Brouwer 1953, 1954). When root growth stops the increased suberization decreases the effective length of absorbing root, and thus the capacity to take up water varies with season and is likely to be least in winter. The inability to satisfy water deficits overnight has often been attributed to root resistance (e.g. Sands & Rutter 1958). Consequently the water deficit at dawn increases during the winter to a maximum before the onset of the new season's growth (Table 6.1). Those species which show some root growth in winter (e.g. *Erica cinerea* and *E. tetralix* which are adapted to mild winters) show lower deficits at dawn than species which show a more definite winter dormancy (e.g. *Calluna vulgaris* which has a more continental distribution).

Root resistances vary with transpiration stress (Brouwer 1953; Tinklin & Weatherley 1966) and plants which cannot fully satisfy

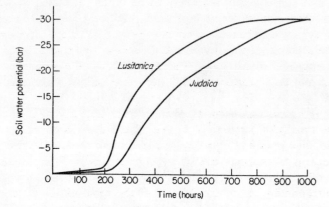

FIGURE 6.10. Depletion of soil water by two subspecies of *Dactylis glomerata* (after McKell *et al.* 1960).

Spp. *judaica* comes from the semi-arid climate of Israel, ssp. *Lusitanica* from the more oceanic climate of Portugal. The subspecies differ both in vegetative form and in rooting frequency (Perrier *et al.* 1961).

their water deficits overnight still transpire, without large increases in water deficit, during the following day. This variation of root resistance with transpirational stress suggests that the simple catenary treatment of Figure 6.7 is an oversimplification.

The situation is more complex in the field as root frequency and distribution, the physical nature of the soil, and amounts of soil water all vary with depth. The effects of soil water and root frequency and distribution in each soil horizon must be integrated to provide a complete picture of the influence of roots on water uptake. The upper layers of the soil usually have an abundance of absorbing

TABLE 6.1. Water deficits (%) at dawn during the dormant season (Aberdeen 57°N). (Data of Bannister 1964c).

date	*Calluna vulgaris*	*Erica cinerea*	*Erica tetralix*
22.9.62	10·0	7·4	1·9
31.10.61	11·5	13·1	9·3
11.2.62	20·5	13·8	11·8
30.4.62	19·7	12·5	5·5
24.5.61	28·4	13·6	14·0

Dates are arranged with regard to time of year and are therefore not in strict chronological sequence.

roots and consequently are likely to be rapidly depleted of water. Plants which exploit deeper layers are able to take up and transpire water, even when the surface layers are dry. Some desert species can tap the subterranean water table and become 'water spenders' in an apparently arid environment. Such species lower the water table yet further and their eradication aids the conservation of water in arid zones.

The few roots that occur at greater depths have a significant effect on the uptake of water only if they have a low resistance per unit length. This is true of most roots but the fine roots of herbaceous species and grasses may not meet this requirement (Wind 1955).

Finally, the soil-plant system is not static. Water can move from the surrounding soil to the vicinity of the roots although rates of movement in dry soil are very slow. Plant roots also grow through the soil and their rate of growth in dry soils probably allows them to exploit new sources of water before soil water has had time to move to the roots (Newman 1966).

6.3.6 Water transport between
the stem and the leaf

The major resistances to water movement within the plant are located in the root and leaf, other plant resistances are small (Tinklin & Weatherley 1966; Jensen *et al.* 1961). Consequently, internal resistances are unlikely to influence the water potentials developed in most herbaceous plants. In shrubs and trees, both the resistance of the conducting vessels and the height of the water column can contribute to decreased water potentials in the plant. These internal resistances have generally been overlooked, particularly in comparative ecological studies (however cf. Huber 1956). Nevertheless shoots of *Erica cinerea* have a lower 'resistivity' to water movement than those of *Calluna vulgaris* (Haines 1928) whilst *Erica tetralix* has a lower resistance to internal water transport than *Calluna* (Firbas 1931). The water deficits of these species in the field (Bannister 1964c) are in accord with their relative conductivities and *Calluna* generally shows the highest deficits, although it must be emphasized that these deficits are the resultant of all resistances in the system and of the relationship between water deficit and water potential for each species (6.4.2).

The resistance of the stems of coniferous trees to water uptake is

several times greater than that of deciduous trees (Huber 1956) and
the water deficits in deciduous trees are usually less than those in
conifers (Jarvis & Jarvis 1963c). Ring porous woods (e.g. ash) are
less resistant to water transport than diffuse porous woods (e.g.
sycamore and willow, Peel 1965).

Woody roots and vines have low internal resistances (Huber 1956).
The rate of water movement within the plant is influenced by its
internal resistance. In conifers and other evergreen trees, the maximum rates of movement are in the order of 0.4 mm s^{-1} (1.5 m
h^{-1}) while deciduous trees may exhibit rates of up to 12 mm s^{-1}.
The highest rates are found in lianas (40 mm s^{-1}) although
rates in herbaceous plants may be as high as 18 mm s^{-1} (Huber
1956).

6.4   TISSUE WATER RELATIONS

6.4.1   *The measurement of the*
water status of tissues

In many instances the water relations of plants are analysed by
measuring water contents (i.e. the difference between fresh, w_f, and
dry, w_d, weights) and expressing them as a percentage of fresh or
dry weight. Similar expressions are used for soil moisture contents
but, whereas soil dry weights remain relatively constant, plant dry
weights vary both seasonally and diurnally. Consequently, the water
content per unit dry weight shows an annual peak in spring or early
summer when shoots are young and succulent and a minimum in
hardened, mature shoots in winter (Figure 6.11). Diurnal variations
in dry weight are caused by the manufacture and translocation of
photosynthate. Thus, if moisture contents are measured on a fresh
or dry weight basis, apparent changes in moisture status can be produced merely by changes in dry weight.

Relative water content (the relative turgidity of Weatherley 1950)
and its complement, water deficit (Stocker 1929) are theoretically
independent of changes in dry weight. They are derived from
measures of fresh (w_f) and dry (w_d) weight expressed as a proportion
(usually percentage) of the water content of tissue which has been
allowed to become resaturated with water (($w_s - w_d$), where w_s is the
resaturated weight).

Relative water content, $W_R = \dfrac{100\,(w_f - w_d)}{(w_s - w_d)}$

and   water deficit, $W_D = \dfrac{100\,(w_s - w_f)}{(w_s - w_d)} = 100 - W_R$

Water deficits and relative water contents are more accurate measures of changes in plant water status than expressions on a fresh or dry weight basis (Figure 6.11). However, there is no unique relationship between water deficit and water potential and care must be taken when making comparisons both within and between species. A high water content may be indicative of high potential in one species or

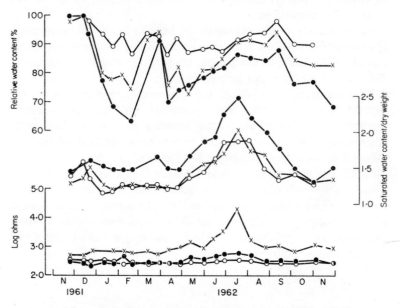

FIGURE 6.11. Annual course of water relations in *Calluna vulgaris* (•), *Erica tetralix* (o) and *E. cinerea* (x) (Bannister 1964c). (Sands of Forvie, Aberdeenshire).

Above: Course of relative water content. Note the low winter values and the amelioration during the growing season.

Centre: Saturated water content per unit dry weight. This follows the annual pattern of shoot extension.

Below: Resistance of soil moisture units. High resistances denote dry soils. Note the higher resistances in the site supporting *Erica cinerea* and the absence of any apparent effect of soil drought upon water relations of this species.

at a particular time of year, but the same water content may be associated with a lower water potential in a different species or at another time of year (6.4.2).

Tissue water potentials can be measured by equilibrating plant material in or above solutions of known osmotic potential. The water potential is estimated from the osmotic potential of the solution (or the water potential of the vapour above it) which causes no change in the water status of the tissue. The critical solution may be found by measuring characteristics of the tissue (e.g. Kreeb 1960), of the solution (e.g. Gaff & Carr 1964, Hellmuth & Grieve 1969) or of the vapour (e.g. Spanner 1951, Richards & Ogata 1958). In each case the critical point is where no change occurs in the measured characteristic.

More recently, water potential has been estimated by applying a positive pressure to a leaf or shoot and finding the force needed to extrude water from its cut petiole or stem (Scholander *et al.* 1965). This 'pressure bomb' can be used in the field.

The osmotic potential of cell sap may be measured by some of the techniques used to assess water potential, but the most common method is by depression of the freezing point (e.g. Ramsay & Brown 1955, Walter 1963). The well-known 'method of limiting plasmolysis' (Crafts *et al.* 1949) is an indirect method which relies on finding a point where turgor pressure is just zero and thus the osmotic potential of the cell sap is equivalent that of the external solution. This section has merely outlined the principles behind some of the techniques for measuring water status of plant tissues. Further details can be found elsewhere (e.g. Bannister 1976) and by reference to specialized literature.

6.4.2   *The relationship between water*
deficits and water potential

The relationship between soil moisture tensions and moisture contents has long been used to characterize the water relations of different soils. Thus, organic soils show a small change in tension for a relatively large change in water content leading to a large difference in tension (cf. Figure 5.3). It is possible to characterize plant tissues in the same way. An early example is provided by a comparison of privet with tomato (Weatherley & Slatyer 1957). An identical change in water content causes a large change in water potential in tomato whilst the more xerophytic privet shows a smaller change in water potential and *Acacia aneura* which is yet more xerophytic, shows an even smaller change in water potential (Slatyer 1960).

These and similar observations (e.g. Jarvis. & Jarvis 1963c) indicate that plants adapted to drier habitats show the shallower type of curve. This conclusion is borne out by studies of closely related species in less extreme climates. In Britain, *Erica cinerea* is typical of drier heathland whilst *Calluna vulgaris* (heather) is found on the moister soils. *Erica cinerea* shows a smaller reduction in water content for a given decrease in water potential than does *Calluna* (Figure 6.12). Consequently, *Calluna* would be expected to show a lower water content than *Erica* when they grow intermingled on the same site and are subjected to the same soil water potential. This does occur (Figure 6.12) and, moreover, *Erica* shows smaller reductions in water content than *Calluna* in drying soil. Similar relationships between tissue water contents, water potential and soil moisture are shown in comparative studies of spruce, pine, birch and aspen (Jarvis & Jarvis 1963c). Those species which show the least change in tissue water content for a given change in water potential

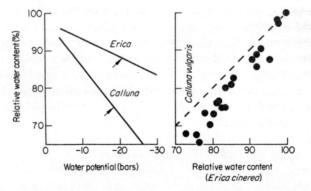

FIGURE 6.12. Relationship of relative water content ot water potential in shoots of *Erica cinerea* and *Calluna vulgaris* (simplified from Bannister 1971) and the correlation of the water contents of the two species from a mixed stand in the field (Bannister 1963).

The dashed line indicates equality of water contents (right) and the arrows the point of stomatal closure (Bannister 1964a). Note that the deviation of the relative water content/water potential curves is in accord with differences in field relative water contents and that the point of stomatal closure occurs at a similar water potential despite the wide discrepancy in water contents.

generally show the highest water contents in the field and are more resistant to the induction of water deficits on drying soils.

Stomatal closure and damage due to desiccation are more likely to be due to reductions in water content than to low water potentials (Jarvis & Jarvis 1963c). Thus, xerophytic species with the shallower type of curve may be able to resist desiccation and maintain open stomata (for example, *Erica* is able to maintain high transpiration rates over a wide range of soil moisture contents, in contrast to *Calluna*—Figure 6.8). However, low water potentials may be damaging in themselves and it has been suggested that species with the shallower type of curve may show a marked reduction of growth with increasing water stress (Jarvis & Jarvis 1963b) so that desiccation resistance and growth rate may be inversely correlated (Jarvis 1963). On the other hand, large reductions in water content cause stomatal closure, prevent the uptake of carbon dioxide and consequently inhibit growth. Compensatory mechanisms also occur so that *Erica cinerea* exhibits stomatal closure at a higher relative water content than *Calluna* (Table 6.3); thus *Erica* is shielded from the effects of low water potentials and *Calluna* is protected from premature stomatal closure.

The relationship between water potential and water content is not as constant within a species as at first thought (cf. Slatyer 1960). Different relationships have been produced in plants subjected to different soil moisture regimes or grown in solutions of various osmotic potentials (Jarvis & Jarvis 1963a,c). The relationship also varies with the age of the tissue (Knipling 1967) and with the habitat in which the plant grows (Knipling 1967, Bannister 1971). In all these cases the less steep curves are produced in the more extreme conditions—in response to the driest regimes and lowest osmotic potentials, in exposed habitats and in older tissues.

Osmotic potentials are also related to water deficits and have long been used as an indication of water stress within the plant, thus plant production has been shown to decrease exponentially with osmotic stress (Kreeb 1963). It has been argued that the degree of hydration of the protoplast is more important for the proper functioning of the plant than the overall water balance. These ideas are embodied in the concept of '*Hydratur*' (hydrature, Walter 1931, 1963) which may be considered as a protoplasmic 'relative humidity' that is directly related to osmotic potential. Thus, a cell with an osmotic potential of -20 bar at $20°C$ has a hydrature of 98·5 per

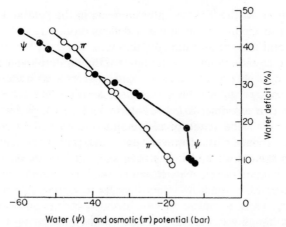

FIGURE 6.13. Interrelationships of water deficit, water potential and osmotic potential in *Acacia craspedocarpa* (after Hellmuth 1969).

Points represent values for different seasons of the year. Note the point of zero turgor ($\psi = \pi$) at a water deficit of 33 per cent and the 'negative turgor' ($\psi > \pi$) at greater water deficits.

cent. Theoretically, the hydrature of a cell varies with the location of measurement (Slatyer 1967). It is difficult to separate effects due to water potential from those due solely to osmotic potential as they are closely related and usually decrease together, because they are both related to tissue water contents (Figure 6.13). At higher water contents the osmotic potential is lower than the water potential as there is a positive turgor pressure, they are equal at a point of zero turgor while at very low water contents the water potential may be exceeded by the osmotic potential. This last condition is often interpreted as one of negative turgor, which implies an inward collapse of the cell walls. If there is no negative turgor and no evolution of bubbles of vapour, then the water in the cells must be under considerable tension or matric potentials must be involved (Slatyer 1960, 1967; Hellmuth 1969).

6.4.3 Water deficits and stomatal closure

The most common ecological interpretation of stomatal closure is that it affords the plant protection from excessive water loss and consequently rapid and effective closure is an adaptation against drought. This sort of reaction undoubtedly occurs and is shown by the coastal saltbush of semi-arid Western Australia (*Rhagodia baccata*) which shows a stable water balance after a deficit of about 31 per cent has been reached (Hellmuth 1968). However, not all plants of dry situations show similar reactions—in the same situation *Acacia craspedocarpa* shows no control of water loss (Hellmuth 1969) and the steppe grass (*Festuca dominii*) shows a similar pattern of transpiration on irrigated and unirrigated plots (Rychnovská & Květ 1963). These paradoxical reactions may be explained by the fact that stomatal closure limits all gaseous exchanges and that control of water loss must also result in a reduction of the intake of carbon dioxide (Figure 6.14). Theoretically, partially closed stomata are more resistant to water loss than they are to carbon dioxide uptake; however, when complete closure occurs, carbon dioxide uptake ceases although water is still lost through the cuticle. When cuticular losses are taken into account, carbon dioxide intake may be reduced

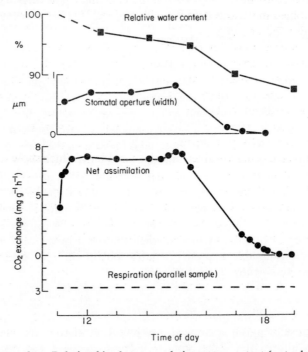

FIGURE 6.14. Relationships between relative water content (water loss), stomatal aperture, photosynthesis and respiration in *Betula pendula* (based on Pisek & Winkler 1956).

at least as much as water loss; this explains the results of workers who have found that the restriction of water loss and reduction of photosynthesis follow a similar course (e.g. Brix 1962). Thus it would seem that species from dry habitats may forego a degree of stomatal protection from desiccation for the sake of continued assimilation.

Other plants show metabolic adaptations which allow them to continue photosynthesis and yet control transpiration by stomatal closure. The efficient fixation of carbon dioxide by plants with a C_4 metabolism (3.2.1) allows them to reduce internal carbon dioxide concentrations to zero when stomata are closed and maintain a steep gradient across the leaf epidermis when they are almost closed. Consequently millet (C_4) transpires about 300 g of water for every gramme of dry matter produced whilst rye (C_3) uses more than 600 g g^{-1} (Black 1971). Plants with crassulacean acid metabolism can close their stomata during the day, and thus restrict water loss, and take up carbon dioxide through open stomata at night. The carbon dioxide is combined in various organic acids and released for photosynthesis the following day (3.3.1).

The interactions between stomatal closure, water conservation and photosynthesis make a study of the effects of water deficits upon stomatal closure of considerable ecological interest. The hydro-active closure of stomata has been examined from an ecological standpoint (Pisek & Berger 1938, Pisek & Winkler, 1953, 1956) as has the decline of transpiration in cut shoots (Hygen 1953).

Some results for water deficits and, where possible, water potentials at stomatal closure are given in Table 6.2. Plants which show stomatal closure at high relative water contents are either typical of moist habitats (e.g. *Populus tremula, Erica tetralix*) or of drier situations if the water content at closure represents a low water potential (e.g. *Erica cinerea*). Within species, samples from moister habitats tend to have higher water contents at closure than those from drier sites (e.g. *Calluna vulgaris, Vaccinium vitis-idaea*). Species which show closure at low water contents may avoid moisture stress if they normally grow in wet soils (e.g. *Vaccinium uliginosum*); otherwise they may be subjected to desiccation (e.g. spruce and pine). Spruce and pine are more tolerant of desiccation than birch and aspen and samples of *Vaccinium vitis-idaea* from dry sites are more tolerant of desiccation than samples from moister sites (see Table 6.5) while *Calluna* is generally more capable of recovering from large water deficits than is *Erica cinerea* (Bannister 1970, 1971).

TABLE 6.2. Relative water contents and water potentials (in brackets) at
stomatal closure in various species.

Trees	% relative water content (water potential in bars)	
Populus tremula[1]	90	(−5)
Betula pendula[1]	80	(−10)
Pinus sylvestris[1]	80	(−15)
Picea abies[1]	71	(−37)
P. abies (sun)[2]	84	
P. abies (shade)[2]	87	
Fagus sylvatica[2]	85	
Quercus robur[2]	87	
Shrubs and dwarf shrubs		
Calluna vulgaris[3]	75	(−18)
C. vulgaris (wet site)[4]	76	
C. vulgaris (dry site)[4]	68	
Erica tetralix[4]	90	
E. cinerea[3]	88	(−21)
Vaccinium myrtillus[3]	79	(−23)
V. vitis-idaea (moist site)[5]	75	
V. vitis-idaea (dry site)[5]	60	
V. uliginosum[5]	63	
Rhagodia baccata[7]	69	(−32)
Herbs		
Asarum europaeum[2]	85	
Stachys recta[2]	68	
Convolvulus arvensis[2]	69	
Potentilla rupestris[6]	90	
P. rupestris (chlorotic)[6]	57	
Lathyrus montanus[6]	90	
L. montanus (chlorotic)[6]	50	
Hypericum hirsutum[6]	84	
H. hirsutum (chlorotic)[6]	79	

References: 1. Jarvis & Jarvis 1963d; 2. Pisek & Winkler; 3. Bannister 1971;
4. Bannister 1964a; 5. Polwart 1970; 6. Hutchinson 1970; 7. Hellmuth 1968.

Ericaceous dwarf shrubs from shaded habitats show a more rapid
stomatal closure and are more susceptible to desiccation than those
from open habitats (Bannister 1971). Thus it seems that species with
low relative water contents at stomatal closure are often resistant to
desiccation, and are adapted so that they maintain open stomata
over a wide range of water deficits.

However, where low relative water contents at stomatal closure are combined with poor desiccation tolerance, plants may be made susceptible to drought. This occurs in chlorotic plants (Table 6.2; Hutchinson 1970, 1971) and may contribute to their elimination from dry calcareous habitats.

6.4.4   Water relations and physiological processes

Plants suffering water deficits often show a reduction or cessation of growth (Figure 6.15). Usually, plants which show small reductions in growth in dry regimes are considered to be adapted to such conditions in nature. From the data of Figure 6.15 it appears that aspen is best adapted to growth in dry soils and spruce the least, although the investigation was concerned with mixed stands of all four species. Ecological experience suggests that aspen is more likely to be found in wetter habitats than birch, spruce or pine but is not particularly common in dry habitats. Aspen also shows the highest relative growth rate in the wettest treatment which appears to inhibit growth of the other species. However, aspen is the least drought tolerant of the four species and closes its stomata in response to the smallest decrease in either water content or water potential and it therefore seems that, in this instance, the ability to survive drought is more important than the ability to grow well at moderate soil moisture tensions. There is not always an inverse correlation between the ability to grow well under moderate soil moisture tensions and drought resistance. This makes ecological interpretation easier. For example, the growth of dogwood (*Thelycrania sanguinea*) is less affected by increasing soil moisture tension than that of bird cherry (*Prunus padus*) and dogwood is also somewhat more drought resistant (Jarvis 1963). Bird cherry is typical of north-western Britain and is confined to cooler, moister slopes of a northerly aspect in the southern part of its range, whilst the dogwood is a plant of south-eastern Britain that is confined mostly to warmer and drier, south facing slopes in the northern part of its range. In this case, the species showing the least reduction of growth and the greatest drought resistance is most typical of the drier habitats. Long-term experiments on growth in different soil moisture regimes may be complicated by changes in leaf area during the experiment. Leaf areas are generally less in plants grown on dry regimes and this reduction may contribute to reductions in growth as well as any

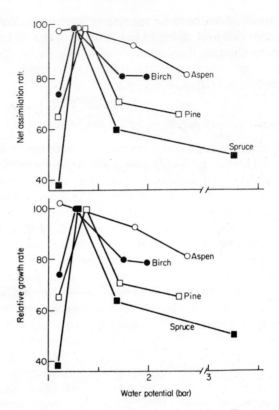

FIGURE 6.15. Growth of tree seedlings in relation to soil water potential. Both net assimilation rate and relative growth rate are expressed as a percentage of that in the second wettest regime. The absolute values in that regime are, (for the sequence aspen, birch, pine, spruce) 22·2, 30·9, 22·7, 18·9 g m^{-2} wk^{-1} (NAR) and 502, 513, 81, 34 mg g^{-1} wk^{-1}. (After Jarvis & Jarvis 1963b.)

direct effects of water deficits upon photosynthesis. The effect of changes of leaf area can be separated by classical growth analysis if the net assimilation rate (NAR), or unit leaf rate (ULR), is also estimated and in drought-adapted plants reductions in leaf area may be offset by increases in NAR, as in C$_4$ plants (3.3.1).

Soil moisture stresses usually affect plant growth by influencing the water deficits produced in plant tissues. In higher plants, water deficits often cause stomatal closure and a consequent reduction in photosynthesis; but photosynthesis is also reduced by water deficits in lower plants which have no stomata (Figure 6.16). There must, therefore, also be direct effects of low water contents and potentials upon photosynthesis.

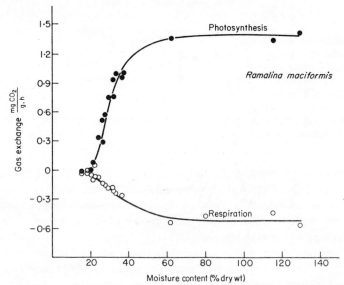

FIGURE 6.16. The relationship of photosynthesis and respiration to thallus water content in the desert lichen, *Ramalina maciformis* (Lange 1969).

Respiration may be little affected by small changes in water content (cf. Figures 6.14, 6.16) but larger reductions may cause increased respiration before an ultimate decline (Brix 1962). The combination of decreased photosynthesis and increased respiration in plants suffering from water deficits results in increased carbon dioxide concentrations in the leaf mesophyll, which, given the sensitivity of the stomatal apparatus to carbon dioxide, may hasten stomatal closure (Meidner & Mansfield 1968).

In lower plants, which may be adapted to a diurnal cycle of hydration, the net production of carbon dioxide is influenced by the sensitivity of respiration rates to temperature. Thus, desert lichens show carbon dioxide assimilation when they are naturally wetted by dew in the cool of the early morning but release large quantities of carbon dioxide if they are wetted artificially in the middle of the day (Table 6.3).

TABLE 6.3. Maximum rates of CO_2 exchange in the desert lichen, *Ramalina maciformis*, after natural and artificially induced water uptake at different times of day (data of Lange 1969).

	time	temperature (°C)	light intensity (k lux)	CO_2 exchange (mg g(dry wt.)$^{-1}$ h^{-1})
natural absorption of water vapour	6·45	21	25	0·30
natural wetting by dew	7·15	21	30	1·57
artificial wetting	13·05	39	100	−0·30
artificial wetting	15·45	19	25	0·40

Positive values of CO_2 exchange indicate net photosynthesis, negative values net respiration.

Growth, photosynthesis and respiration are all affected by water stresses in the plant and are obviously interconnected, both through stomatal movements and carbohydrate metabolism. However, water deficits directly affect the increase in size of the plant or its component parts through the action of turgor pressure in cell expansion. Decreased turgor pressure is more inhibitory to cell expansion than is any associated decrease in osmotic potential (Ordin 1958, 1960). Many plants appear to be able to maintain turgor pressures by osmotic compensation (i.e. the uptake of salts or the mobilization of reserves as soluble sugars) over a wide range of water potentials. The osmotic compensation may be more or less complete in halophytic plants (e.g. Black 1960; Hellmuth 1968) and even crop plants (Eaton 1942) but where it is incomplete (e.g. Scholander *et al.* 1962) the plants must also suffer from the effects of decreased water availability (i.e. physiological drought).

Water stress also has adverse affects upon other processes, cell division appears to be less severely affected than cell expansion (Gardner & Niemann 1964) but growth may be further inhibited by changes in carbohydrate metabolism (Eaton & Ergle 1948), nitrogen metabolism (Shah and Loomis 1965) & possibly in the production of growth substances (Larson 1964) and the translocation of materials (Roberts 1964). The fact that many of these processes are controlled by enzymes suggests that water stress may affect proteins directly. Various theories exist to explain the results of dehydration upon proteins, these include the sulphydryl/disulphide hypothesis of Levitt (1962), changes in the conformation of protein molecules (Gaff 1966) and in the structure of their hydration lattices (Slatyer 1967).

In conclusion it appears that a plant's reaction to increased moisture stress is complex but that most of the physiological changes are brought about by the changes in the water relations of the tissues of the plant itself. Stomatal movement, cell turgor and direct metabolic effects all influence the growth and hence the ecological response of the plant. Many ecological studies are concerned only with the effect of decreased soil moisture levels on some readily measured parameter such as growth—better interpretations could be made from the concurrent examination of plant water status and specific physiological processes.

6.4.5 The water relations of groups of plants

It should be clear from the preceding sections that there is no unique strategy in plant water relations—even for different species in the same habitat. Nevertheless, broad ecological generalizations can be made. Plants from moist habitats generally show lower and less variable water deficits, and higher water and osmotic potentials than plants from more extreme environments (Table 6.4). The lowest values are found in plants from arid zones and in halophytes—these species often have to endure low water potentials in their environment—but low values are also found in trees which develop low water potentials in their topmost twigs because of the continuation of the transpiration stream beyond the height of water that could be supported by atmospheric pressure alone.

TABLE 6.4. Range of water potentials[1] and osmotic potentials[2] in various groups of plants.

	water potential	osmotic potential
Water plants	$-2--12$	$-4--16$
Deciduous trees	$-9--23$	$-10--28$
Coniferous trees	$-5--60$	$-10--60$
Arid shrubs	$-22--80$	$-12--60$
Halophytes	$-24--60$	$-12--50$

[1]Based on data of Scholander et al. (1965) & Walter (1960)[2]. The results are not strictly comparable as they were made on different lots of plants. Except where 'negative' turgors occur (as in arid shrubs) water potentials would be expected to be higher than osmotic potentials.

There are two broad systems of water balance (cf. Stocker 1956). In the first, isohydric, type transpirational control ensures no great diurnal fluctuations in water deficit or osmotic potentials. The second, anisohydric, pattern is shown by plants which show little control of transpiration and tolerate the development of large deficits within their tissues (Figure 6.17). As has been seen, the presence of these two types is a consequence of the conflicting demands of water conservation and continued photosynthesis.

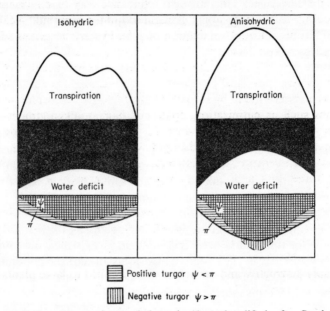

FIGURE 6.17. Types of water balance in plants (modified, after Stocker 1956).

The isohydric type controls all aspects of its water relations, usually through stomatal closure, and maintains an equable balance. The aniso-hydric type shows little control and allows large fluctuations in its water relations, even to the extent of developing 'negative' turgor.

6.5   DROUGHT RESISTANCE

6.5.1   *Terminology and measurement*

Drought resistance is a general term for the ability of plants to survive the consequences of drought. There is no universal way by which this is achieved and in consequence the different components of drought resistance have been classified in different ways by different authors. One classification which seems to have gained a more general acceptance (cf. Parker 1969) is that due to Levitt, Sullivan & Krull (1960). In this system drought resistance is retained as a general term but the ability of tissues to withstand low water potentials is termed drought tolerance whilst the various methods which enable plants to escape low water potentials being developed constitute drought avoidance. Thus drought avoidance may be considered as an ecological response whilst drought tolerance is physiological and protoplasmic. Drought avoidance may be by various means; desert ephemerals avoid drought by completing their life cycle when moisture conditions are favourable, deep rooted species may tap subterranean water and be water spenders (drought evaders) whilst others conserve water by stomatal closure and xeromorphic modifications such as cutinization, epidermal hairs, sunken stomata and water storage (drought endurers). In a natural situation the ability of a plant to resist drought is compounded of both avoidance and tolerance; drought resistant plants both exercise control over water loss and have a degree of tolerance of high water deficits.

The most valid assessment of drought resistance would undoubtedly be by monitoring the survival of representative numbers of plants subjected to controlled levels of soil and atmospheric drought. However, such methods are often impractical and invariably destructive and cannot be easily applied to large plants such as mature trees, although comparative studies of the survival of trees have been made during times of severe natural drought. The results of such studies have been collected together and tabulated (Parker 1969). For most practical purposes, some more rapid assessment of drought resistance is needed and isolated plant parts, particularly shoots and leaves, have often been used, either by allowing dehydration for different lengths of time (e.g. Bannister 1970) or by using environments of different evaporation potential (e.g. Jarvis & Jarvis 1963c). Drought avoidance may be measured by monitoring the water deficits induced during desiccation whilst tolerance may be assessed by determining the water status of the material at a point where a critical level of damage is induced or by examining the ability of the tissues to make up their induced deficits. Further details of technique are to be found elsewhere (e.g. Bannister 1976).

The changes in tolerance that occur are probably caused by a number of mechanisms including changes in the structure and hydration of proteins (Slatyer 1967, Levitt 1972), changes in the relationship between water content and water potential (6.4.2) and alterations in the amount of 'bound' water that is unable to take part in normal cellular water exchanges.

6.5.2   Ecological aspects

The drought resistance of a plant varies with its previous history and stage of growth as well as with conditions prevailing at the time of measurement. Avoidance and tolerance mechanisms may also vary independently of one another. Thus, deciduous species are highly resistant to the formation of water deficits in winter, whereas evergreen species may be most resistant in summer (Figure 6.18). On the other hand, older tissues are often more tolerant of water deficits than younger tissues. Desiccation tolerance is therefore lowest at the beginning of a new growing season, although species differ in their rate of hardening during summer (Figure 6.18). Consequently, different patterns of drought resistance emerge and these are illustrated by the species in Figure 6.18, where *Vaccinium* and *Calluna* are most resistant in winter (with tolerance playing a large part in the resistance of *Calluna* and avoidance being most important in *Vaccinium*) while *Erica* is not particularly resistant to winter drought, (but the avoidance and tolerance mechanisms reinforce each other in summer). *Erica* is confined to oceanic areas with mild, wet, winters whereas the other two species have a greater continental and boreal distribution and appear to be adapted to more severe winters.

Drought resistance studies have been used (Rychnovská 1965) to differentiate between mesophytic grasses of a somewhat oceanic distribution (*Corynephorus canescens, Bromus erectus*) and xerophytic grasses of a continental distribution (*Stipa capillata, Festuca dominii*). The latter grasses are more able to regain their initial water contents after desiccation. Drought tolerance also differs within a species. For example, *Vaccinium vitis-idaea* from exposed turf shows a much greater tolerance of deficits than material of the same species from a wet ledge (Table 6.5). Plants in different physiological states may also vary in their drought resistance; chlorotic plants show little resistance to the induction of deficits (Hutchinson 1970a) and a reduced ability to recover from induced deficits (Hutchinson 1970b).

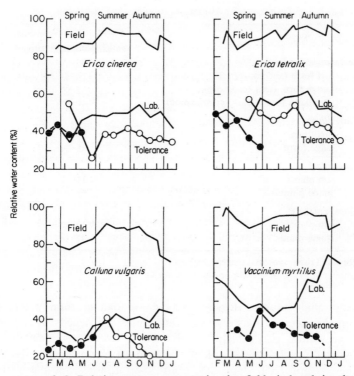

FIGURE 6.18. Relative water contents in the field, induced in the laboratory (mean values) and the relative water content required to induce slight (20 per cent) damage (tolerance). Based on Bannister 1970.

In *Erica* and *Calluna* field and laboratory water contents tend to be low in winter and spring and higher in summer and autumn (drought avoidance). In *Vaccinium* the leafless stems resist the formation of water deficits in winter. In *Erica* the maximum tolerance appears to be in spring and summer whereas in *Calluna* and *Vaccinium* it is in autumn and winter.

Open and filled circles differentiate between current and previous season's growth.

Shaded dwarf shrubs are also less able to resist the formation of deficits and more intolerant of low water contents than plants from the open (Bannister 1971). In both these instances the reduced drought resistance may explain the gradual elimination of the affected species from calcareous and shaded habitats respectively.

In the field, lethal water contents are rarely attained as plants are usually well adapted to the environments in which they grow.

TABLE 6.5. Differences in desiccation tolerance within species.

Species	relative water contents causing damage (range)
Vaccinium vitis idaea[1]	
turf form (xeromorphic)	16–51
Ledge form (less xeromorphic)	35–60
woodland form (least xeromorphic)	40–55
V. myrtillus[1]	
turf form	15–36
ledge form	25–39
V. myrtillus[2]	
sun plants	<43–42
shade plants	55–69
Calluna vulgaris[2]	
sun plants	<26–50
shade plants	31–58

[1] Polwart (1970, seasonal extreme values.
[2] Bannister (1971), mean winter and summer values.

Exceptional summer drought may cause damage and plants of temperate regions may show extreme moisture deficits at the end of the winter, particularly if soil water is frozen and unavailable (Figure 6.11). In general, plants from moist environments are less likely to approach damaging limits than those from more arid situations (Table 6.6).

TABLE 6.6. Field water deficits expressed as a percentage of the water deficit required to induce slight damage. (Adapted from Larcher 1973a and incorporating data of Bannister 1964c, 1970, 1971).

Vegetation type	Field W_d/sublethal W_d(%)
Woodland herbs	6–25
Deciduous trees and shrubs	10–64
Oceanic dwarf shrubs	15–40 (88)
Meadow grasses	19–74
Cultivated plants	44–71
Herbs of dry habitats	43–96
Steppe grasses	50–80 (108)
Mediterranean trees and shrubs	47–90 (105)

Field water deficits are the average maximal values. Figures in brackets represent extremes. It is apparent that plants of dry habitats are more likely to approach limits that impose damage than plants from moist habitats.

6.6 EXCESS OF WATER

An excess of water is almost as inimical to plant growth as a deficiency. Water displaces air in the soil and waterlogged soils are deficient in oxygen and induce anerobiosis in roots. In contrast to drought resistance, plant resistances to water excess are mostly avoidance mechanisms. The most common method of avoidance is by the provision of internal air spaces, typical of almost every hydrophyte, which may interconnect the leaves and roots of emergent plants. Many plants evidently transport oxygen from leaves to root; these include grasses of wet places such as *Molinia caerulea* (Webster 1962b) and cultivated cereals such as rice (Barber *et al.* 1962) and even maize (Jensen *et al.* 1967). Pineapple, with its crassulacean acid metabolism, may have closed stomata during the day while it is photosynthesizing and consequently shows a large export of oxygen to its roots (Ekern 1965). Most plants produce the bulk of their absorbing roots above a permanent water table in the soil and this also avoids anaerobiosis. Physiological avoidance also occurs, as

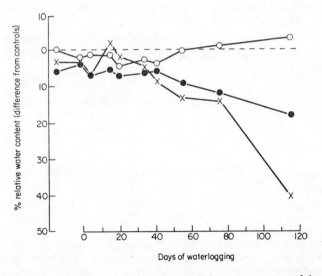

FIGURE 6.19. The difference between the relative water content of shoots from waterlogged plants of *Calluna vulgaris* (●), *Erica cinerea* (X) and *E. tetralix* (○) and that of controls (unwaterlogged). After Bannister (1964d).

plants tolerant of flooding develop alternative biochemical pathways which prevent the accumulation of toxic metabolites in waterlogged plants (Crawford 1972). Plants which occur in waterlogged soils may also have to be tolerant of toxic chemicals which are produced in the reducing conditions that obtain in such soils.

Plants that are killed by waterlogging often appear desiccated. Leaves become yellowed and fall off and plants often show a reduction of transpiration and an increase in water deficit (Figure 6.19). The death of roots could lead to an increased root resistance and a resultant decrease in the uptake of water, although plants often appear moribund before any dramatic increase in water deficits occurs. The yellowing of leaves has been compared with ethylene poisoning (Kramer 1951) and this is not an impossible cause of death. Ionic toxicity may also be responsible and ferrous and manganous ions are likely to accumulate in anaerobic soils (Jones

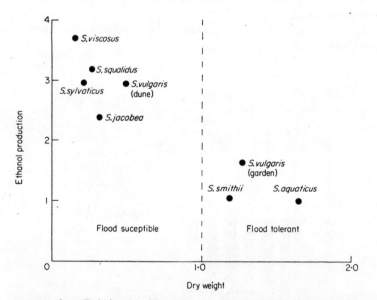

FIGURE 6.20. Relationship between growth and ethanol production in *Senecio* species grown on soils with high and low water tables (data of Crawford 1966).

Both dry matter and ethanol production are expressed as the ratio of the amount produced on flooded soil to that in a soil with a low water table. The increased growth of flood-tolerant species is associated with lower ethanol contents.

& Etherington 1970). Within plants, the accumulation of ethanol (Figure 6.20) is a likely cause of death.

The interplay of these various factors is illustrated by a comparison of *Erica cinerea* and *E. tetralix*. *E. tetralix* is found on sites where the water table is consistently high (Rutter 1955) and its distribution is correlated with the presence of reducing conditions, as indicated by the presence of hydrogen sulphide, in the soil (Webster 1962a). It appears to be tolerant of both waterlogging and reducing conditions. On the other hand *E. cinerea* is typical of drier, freely drained sites (Bannister 1965). If the two species are subjected to artificial waterlogging, *E. cinerea* is rapidly killed and shows reductions in both transpiration and water content while *E. tetralix* survives and its water relations are hardly altered (Bannister 1964c). *E. tetralix* accumulates nontoxic malate in its tissues (Crawford & Tyler 1969) and also accumulates less iron and is more tolerant of artificially applied iron than *E. cinerea* in waterlogged conditions (Jones 1970a,b).

The metabolic avoidance exhibited by flood tolerant plants has been extensively investigated. Intolerant plants accumulate large

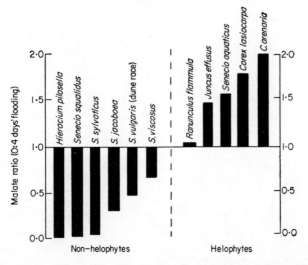

FIGURE 6.21. The malic acid content of species both tolerant (helophytes) and intolerant of flooding (nonhelophytes) expressed as the ratio of the contents before and after four days of flooding. (Crawford & Tyler 1969).

amounts of ethanol whereas tolerant plants accumulate less (Figure 6.20). Differences occur not only between species but also within races of the same species so that *Senecio vulgaris* from a garden soil accumulates proportionately less ethanol than a race from a sand dune (Crawford 1966). Plants which avoid ethanol production often produce malate (Figure 6.21) although some hydrophytes (e.g. *Iris pseudacorus, Nuphar lutea*) accumulate shikimate (Tyler & Crawford 1970). An integrated scheme for the various metabolic alternatives has recently been produced (Crawford 1972).

7: Mineral Nutrition

7.1 INTRODUCTION

7.1.1 *Mineral elements essential to plants*

All plants have certain basic requirements for mineral constituents from the soil. If these requirements are not met then the plant is unable to grow properly. Some elements such as nitrogen, potassium and phosphorus are needed in large amounts and thus form the major components of inorganic fertilizers; substantial but lesser amounts of iron, magnesium, sulphur and calcium are also essential to the growth of plants. These seven are generally termed the major elements. Other elements are needed in very small amounts, indeed, they are often toxic in large quantities, and in commercial fertilizers are usually present in sufficient quantities as contaminants. These include boron, manganese, molybdenum, copper, zinc and chlorine and in addition sodium in some Chenopodiaceae, cobalt in members of Leguminosae which symbiotically fix nitrogen, aluminium (ferns) and silicon (diatoms). These are the minor or trace elements.

In most cases it is relatively easy to appreciate the requirement and utilization of the elements by the plant. Nitrogen is a component of proteins and thus of all enzymes; it is also part of many other cell chemicals such as chlorophyll, nucleic acids, nucleotides, coenzymes and vitamins. Phosphorus is an essential part of the ATP molecule as well as of nucleic acids and phosphorylated sugars and phospholipids. Potassium is often present in large amounts but its function is not so readily seen as that of the other two most abundant minerals; it is concerned in osmo-regulation and is an essential cofactor for many enzyme systems. Calcium as calcium pectate is a constituent of the middle lamella of plant cells and has effects on enzyme systems and cell permeability (Wyn-Jones & Lunt 1967) whilst magnesium is an essential part of the chlorophyll molecule and sulphur is a constituent of some proteins. Other major elements and trace elements are mostly essential to various enzyme systems.

171

The agriculturalist is concerned with maximum yield and thus wishes to optimize the nutrient supply to plants. Under natural conditions, plants from a particular habitat are tolerant of local conditions of nutrient supply and able to persist on soils which, without the addition of fertilizer, would be useless for agriculture. On such soils the addition of complete fertilizer often leads to dominance by a single species and the elimination of diversity. This is well illustrated by the work of Willis (1963) where an addition of complete nutrients in dry dune pasture led to the dominance by the grasses, *Festuca rubra* and *Poa pratensis*, whereas in wet dune slacks *Agrostis stolonifera* became dominant. When only nitrogen and potassium were added to the slack vegetation *Agrostis* was not favoured but improved growth was shown by species of *Carex* and *Juncus*. Such studies as this and others (Bradshaw 1969) and long-term experiments such as the Park Grass experiment (Thurston 1969) show how the relative level of nutrients may determine the species composition of a particular area.

7.1.2   *pH and mineral availability*

The reaction of the soil has a direct effect upon the availability of both essential and nonessential elements in the soil. Many elements are more soluble and thus more available to the plant in acid soils. Their solubility may be increased to such an extent that they become toxic. Aluminium, iron and manganese are more readily extracted below pH 5 and all may be toxic (e.g. Grime & Hodgson 1969, Jones & Etherington 1970). Other elements which also become more available in acid soils include boron, copper and zinc. The six elements mentioned become less available on alkaline soils so that the toxicity that they may exert in acid situations no longer exists and may be replaced by deficiency symptoms. This is particularly true of iron; many plants, particularly those adapted to acid soils become chlorotic and yellow on calcareous soils; these symptoms can be alleviated by the presentation of appropriate iron compounds (Grime & Hodgson 1969).

However, most major elements become less available in acid soils. This is true of nitrogen, phosphorus, potassium, sulphur, magnesium and calcium as well as of trace elements such as molybdenum (Truog 1947). Nitrogen is additionally sensitive in this respect as microbial nitrification occurs only in a narrow pH range around neutrality.

The simple considerations of availability are made more complex by interactions between the supply of a particular element, microorganisms and the requirement of the plant. Thus, phosphorus may be deficient in alkaline, calcareous soils because of its precipitation as insoluble phosphates; in other soils it may be limited because it is rapidly taken up by the microflora (cf. Barber 1969). On the other hand, deficiencies may be alleviated by the enhanced uptake of mycorrhizas (Harley 1969) or a vegetation may consist of plants that are adapted to low phosphorus levels (Clarkson 1967).

7.1.3 pH and plants

Early in the history of plant ecology, plants were classified according to their substrate preferences and for some time controversy raged as to whether the chemical (e.g. Unger 1836) or the physical properties (e.g. Thurmann 1849) of the substrate were more important in determining the distribution of plants. The chemical theory prevailed (e.g. Warming 1896, Schimper 1898) and the soil reaction and its chemical composition became of prime interest. Species were classified with regard to their substrate preferences and so were described as 'silicicolous', 'halophilous', 'nitrophilous', and 'calciphilous'. The presence of lime was, perhaps unfortunately, frequently associated with the absence of acidity and plants became classified as calcicolous and calcifugous (lime-dwelling and lime-fleeing) and resulted in the formulation of a 'calcicole-calcifuge problem'. Strictly, investigation of this problem should compare plants which cannot tolerate lime with those that can or which even have a high requirement for lime. In practice calcicolous and calcifugous plants are limited not only by the presence of lime but also by other factors. These are discussed later (7.3.1).

There are relatively few studies of the relationships between a wide range of plant species and pH, although horticultural lists have long classified plants according to their 'preference' for certain soil reactions. However, a recent, large-scale, survey of grassland in the Sheffield region (Lloyd et al. 1971, Grime & Lloyd 1973) allows reliable frequency-histograms to be constructed (Figure 7.1). Certain species (e.g. Deschampsia flexuosa) have a marked peak at a low pH, whilst others (e.g. Scabiosa columbaria) have a marked peak at high pH. There are many intergradations between these two

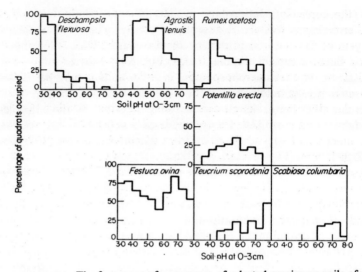

FIGURE 7.I. The frequency of occurrence of selected species on soils of different pH (based on Grime & Hodgson 1969) (see text).

extremes: species such as *Festuca ovina* appear to be largely independent of pH whilst other wide-ranging species may peak in acid soils (e.g. *Agrostis tenuis, Rumex acetosa*) or in soils with a neutral to alkaline pH (e.g. *Teucrium scorodonia*). Other species (e.g. *Potentilla erecta*) are excluded from sites of the highest and lowest pH and show a peak in the mildly acid soils. This study shows that there can be no strict classification into species with a marked preference for acid or alkaline substrata, but that a continuum exists. Much of this continuity is made possible by the great adaptability of species to soil conditions—species may have edaphic ecotypes which may be adapted not only to exotic soils (e.g. those contaminated with heavy metals—Gregory & Bradshaw 1965) but also to a range of 'normal' soils (e.g. Davis & Snaydon 1973). The comparison of ectotypes is a valuable field of study as the number of intrinsic differences between two members of the same species is smaller than between representatives of two quite different species (cf. Grime & Hodgson 1969).

The correlation of pH with plant distribution tempts the ecologist to treat this relationship as one of cause and effect so that variation in soil pH is considered to determine the distribution of plants. In gross terms this is probably correct but there is always the possibility of the converse relationship; i.e. that the distribution of plants determines the soil pH. A general illustration of this is that surface layers of the soil are, in moist temperature climates, usually more acid than the subsoil and that this acidity varies with the plant cover. Thus on the same parent material a planting of coniferous species leads to greater acidification than a planting of deciduous species. A similar situation exists on chalk heathland where heather (*Calluna vulgaris*) and gorse (*Ulex uropaeus*) bushes cause acidification both in superficial layers of the soil and at a greater depth in proportion to their size (Figure 7.2). A slightly different situation has been described for the oceanic grass *Corynephorus canescens* near the eastern limit of its range in Hungary (Rychnovská 1963). No relationship could be found between the distribution of this species and pH, but experiments showed that the roots of the species were capable of changing the pH of unbuffered solutions of a wide range of pH values (3·4–9·42) to a range between pH 5 and 7; solutions of pH 5·8 showed little or no change. In the field *Corynephorus* occupies sandy soils with a low buffering capacity where it is possible for the soil reaction to be adjusted to a value suitable for the plant (Fig. 5.6). Thus a straightforward casual interpretation of the relationship between soil pH and plant distribution is not always

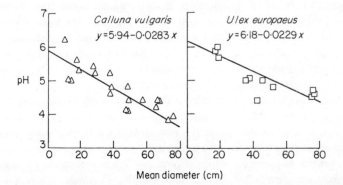

FIGURE 7.2. The pH of the mineral soil under the centre of isolated bushes as a function of the means of their largest and smallest diameters: (a) at 0–1 cm under *Calluna vulgaris*, (b) at 0–1 cm under *Ulex europaeus* (Grubb *et al.* 1969).

The pH of the mineral soil at 0–1 cm depth under the centres of isolated bushes of *Calluna vulgaris* and *Ulex europaeus* as a function of the means of their largest and smallest diameters (Grubb *et al.* 1969).

valid as soil pH may be determined by the plants which grow in a particular site.

7.2 THE INFLUENCE OF MINERAL NUTRITION ON GERMINATION AND ESTABLISHMENT OF SEEDS

There is little evidence to suggest that an external supply of minerals is essential for germinaion. Seeds usually possess substantial reserves which provide sufficient minerals both for germination and establishment. These seed reserves may protect young seedlings from the effects of soil deficiencies and toxicities. Thus, when ryegrass seedlings are used to determine labile phosphorus in the soil (Larsen & Sutton 1963) the seed reserves of phosphorus influence the results for up to six weeks after germination, and the onset of aluminium toxicity in sainfoin can be detected in terms of yield only three–four weeks after germination (Rorison 1969). On the other hand, imbibing seeds will take up mineral ions if they are present in solution and these may affect both germination and radicle extension. Solutions containing approximately 1 mM aluminium depress or delay germination in some grasses, although equivalent amounts of manganese cause stimulation (Hackett 1965). Even lower concentrations (0·1 mM) of aluminium reduce radicle extension in radish and seeds presoaked in aluminium solutions still show an inhibition of radicle extension when transferred to germination pads supplied with pure water (Figure 7.3). In natural situations the soil solution contains a mixture of ions and marked effects are unlikely unless there is a predominance of a particular ion.

The concentration of nutrients does seem to affect the germination and establishment of some species. The rosebay willowherb (*Chamaenerion augustifolium*) shows increased germination with increased nutrient supply, and a similar but less marked effect is seen in *Epilobium montanum*, whilst *Epilobium adenocaulon* and *Tussilago farfara* are unaffected by nutrient supply. *Tussilago* appears to be the best adapted to colonizing soil low in nutrients and *Chamaenerion* would appear to be adapted to the colonization of fertile soils (Myerscough & Whitehead 1966). Species may also be adapted to germinate on soils of a particular chemical composition. Calcareous ecotypes of *Euphorbia thymifolia* germinate best on calcareous soils (Ramakrishnan 1965) although the germination of both calcareous

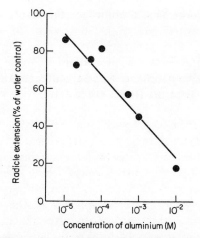

FIGURE 7.3. Effects of pre-soaking radish seeds for 24 hours in dilute solutions of aluminium sulphate upon subsequent radicle extension.

and acidic races of *Melilotus albus* is enhanced by the addition of calcium and phosphorus with the greater stimulation being shown by the acidic race (Ramakrishnan 1968 a,b).

In the field the effects of the chemical properties of soils and those of other physical factors are not readily separated. For example, Rorison (1967) showed that the radicle emergence and subsequent establishment of some calcicolous species on acidic soils was associated with chemical factors that could be alleviated by raising the pH of the soils; whereas calcifugous species could establish themselves on calcareous soils, but that their subsequent poor growth rendered them susceptible to elimination by other factors such as frost and drought. The establishment of foxglove, *Digitalis purpurea*, was influenced by soil texture. Chemical factors, therefore, may not always be responsible for the success or failure of seeds or seedlings on soils of different nutrient status. On the other hand, large amounts of dissolved salts lower the water potential of the soil solution and may reduce the supply of water to the seed; although seeds may be capable of considerable osmotic adjustment and can imbibe water from media with a very low water potential (6.2). Low water potentials are, however, still considered more likely to be responsible for the poor germination of the halophytic grass, *Iva annua*, than ionic toxicities as the seeds show similar responses to a variety of osmotica (Ungar & Hogan 1970). Halophytic species generally show an enhanced ability to germinate in saline media at low temperatures (Figure 7.4). Some, e.g. *Spergularia marina*, *Suaeda depressa*, show some stimulation of germination after pre-soaking in dilute solutions of sodium chloride while seeds of *Spergularia rupicola* germinate well only after soaking in diluted sea-water at low temperature (Binet 1968).

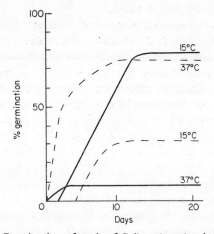

FIGURE 7.4. Germination of seeds of *Salicornia stricta* in rainwater and seawater at 15°C and 37°C (modified and redrawn from Binet 1968). Unbroken line, seawater; broken (dashed) line, rainwater.

After germination, the seedling becomes much more susceptible to external influences. Inhibition of root growth during germination and establishment will render the seedling sensitive to drought and, once the seed reserves are exhausted, it will become dependent upon external sources of minerals and may lose out in a competitive situation.

7.3   NUTRIENT DEFICIENCY

A deficiency of mineral nutrients will limit the distribution of plants in the field. If soils are unable to supply enough of an essential element to a particular species of plant, it may lose competitive vigour and become eliminated from that habitat. In contrast, plants which are adapted to the peculiar conditions of the same site will be able to obtain adequate supplies of the appropriate nutrient. They may achieve this in two ways—either by having a lower requirement for the element or by being able to absorb it efficiently under conditions of low supply. The converse is true of deficient plants—they will either have high requirements or be unable to absorb sufficient quantities of the appropriate element. Thus deficiency to one species may represent sufficiency to another and the following discussion should be read with this in mind.

It follows that symptoms of nutrient deficiency are not commonly seen in areas of established natural or semi-natural vegetation. They may be observed in plants which colonize bare areas and are particularly apparent when plants are grown artificially on alien soils or when continuous cultivation has exhausted mineral reserves in the soil. In the Northern Hemisphere the last glaciation mixed and redistributed soil parent materials over vast areas and did much to replenish nutrient supplies. On the other hand, ancient non-glaciated areas such as South Africa and Australia have been depleted of nutrients over many millenia and soils may be deficient in both major and minor nutrients. In Australia the addition of at least eight nutrient elements (e.g. P, N, Ca, K, Cu, S, Mb, Zn) may be needed to ensure the adequate growth of crops (Grundon 1972).

Many nutrient deficiencies, particularly of minor nutrients, have only been identified after scrupulous experimentation under controlled conditions in the laboratory (Hewitt & Smith 1975) and are unlikely to be detected in the field. Copper and zinc deficiencies come into this category. Copper has been found to be deficient in soils derived from peats (e.g. Mulqueen et al. 1961) and adequate growth of potatoes in peaty soils in Florida is ensured by adding small amounts (approx. $3g\ m^{-2}$) of hydrated copper sulphate (Allison et al. 1927). Zinc deficiencies in fruit trees cause die-back of shoots and abnormal development of leaves and fruit (Alben et al. 1932) and can often be cured simply by driving galvanized nails into the treetrunk.

7.3.1 Deficiency symptoms

The symptoms of nutrient deficiency have been most carefully tabulated for crop plants (Wallace 1961) but comparable symptoms can be observed in wild plants. For example, a large number of plants have been screened for susceptibility to lime chlorosis, which is presumably due to the unavailability of iron in calcareous soils (Hutchinson 1967, Grime and Hodgson 1969), and symptoms of phosphorus deficiency have been identified in *Agrostis* spp (Clarkson 1967).

One of the most obvious signs of nutrient deficiency is a reduction in growth. Stunting may be caused by a deficiency of any element but nitrogen and calcium deficiencies cause severe stunting. The diagnosis of mineral deficiency is usually achieved by combining an assessment of the reduction of growth with other visual symptoms— the two most common being necrosis (death of tissues) and chlorosis (lack of greenness). Deficiencies of nitrogen, sulphur, iron, manganese, or magnesium are all capable of causing chlorosis. In nitrogen deficiency severe stunting is combined with a general chlorosis, particularly of older leaves. Sulphur deficiency causes similar symptoms but growth is better and the leaves may be pale green with lighter areas around the veins. Iron, manganese and magnesium deficiencies produce a mottled chlorosis. Magnesium deficient plants show a marked chlorosis of older leaves although the midribs and main veins may remain green while manganese and iron deficiencies cause a similar chlorosis but of younger leaves. Iron deficiency may also produce small necroses.

In contrast, phosphorus deficient plants are not chlorotic and the young leaves are often dark and bluish-green while older leaves and stems may be reddish. The plants are stunted and slow to produce flowers and fruit. Potassium and boron deficiencies are characterized by necroses. The leaves of potassium deficient plants die at their tips and edges and may be crinkled and distorted prior to death, whereas boron deficiency causes necroses of young roots and shoots. Calcium deficiency rarely occurs in the field. The beneficial effects of liming usually relate to the improvement of soil structure, the reduction of acidity and the alleviation of toxicity rather than any reduction of deficiency. When calcium deficiency does occur the plants are severely stunted with green but distorted leaves.

When visual symptoms occur in the field, it is usually assumed that the soil is deficient in the particular nutrient. There may be confusion when multiple deficiencies occur or when some other agency has produced the symptoms. Mottling may be caused by diseases, particularly by viral infections and various pollutants may cause symptoms similar to those used for the diagnosis of nutrient deficiency (Treshow 1970). Also, the presence of deficiency symptoms in a particular species does not necessarily mean that other species in the vegetation are similarly deficient as plants differ both in their requirements and their capacities to absorb various nutrients.

Agriculturalists have often assessed the mineral status of soils by growing crop plants upon them. If soils are grossly deficient then both crop plants and wild plants will show reduced growth. Proctor (1971a) has used oats for the bioassay of serpentine soils whilst Atkinson (1973) has assayed dune soils by using the weed *Rumex acetosa* which has lower requirements than crop plants and Rorison (1967) has used species with different ecological tolerances for the bioassay of a range of soils (7.2). The improved growth of a test species following the addition of nutrients is often taken as an indication of deficiency (Figure 7.5) although care must be taken in the interpretation of such a response as the addition of nutrients may merely have alleviated toxicity or improved the supply of some other element as a result of changes in soil properties such as pH.

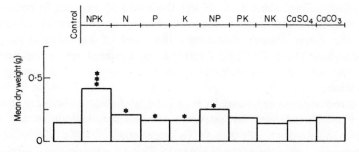

FIGURE 7.5. The response of oats to fertilizer additions on a nutrient deficient serpentine soil from Unst, Shetland (Proctor 1971b).
   Asterisks indicate a significant difference from the unfertilized control soil (* $P < 0.05$, *** $P < 0.001$).

The chemical analysis of plants and soil provides some assessment of nutrient deficiencies, although it is often difficult to relate soil analyses to the availability of particular nutrients to plants and tissue analyses presuppose that there is a 'normal' tissue concentration for each element. The relationship between tissue concentration and yield probably provides a better assessment of deficiency. The yield of deficient plants increases as tissue concentrations increase, except at very low yields where a decrease in concentration may be associated with an increase in growth as nutrients are distributed over a larger volume. Maximum growth is accompanied by an optimum tissue concentration but tissue concentrations may continue to rise above this level without any alteration in yield. This is 'luxury consumption' which is followed by increased concentrations and decreased yield as toxic effects become apparent. 'Normality' is difficult to define and must be associated with the absence of marked deficiency or toxicity.

7.3.2 *Ecological responses to low supplies of nutrients*

Plants which are adapted to low supplies of essential elements will be able to persist on nutrient-poor sites at the expense of species which lack similar adaptations. The elements that have most commonly been studied are nitrogen and phosphorus and these are examined here.

Nitrogen supply

Ecologists have tended to neglect the study of nitrogen in natural communities although nitrogen deficiency is recognized as one of the major factors that limits the yield of agricultural crops (Bradshaw 1969). Ecological interest has centered on such aspects as the insectivorous habit in plants and in symbiotic nitrogen fixation.

Insectivorous plants usually exist in boggy habitats which are short of available nitrogen. In temperate climates they include species of *Drosera* (sundews), *Pinguicula* (butterworts), *Utricularia* (bladderworts) and *Sarracenia* (pitcher plants). There is a tendency to dismiss the influence of the insectivorous habit as unimportant. This view dates from the scepticism of the eminent physiologist, Pfeffer (1877) but there has been sufficient experimental work since to suggest that he was incorrect. As early as 1878, Francis Darwin showed that *Drosera* plants which were fed insects grew better than those that were not and this conclusion was substantiated in plants grown from seed (Busgen 1883). More recently, *Drosera* plants which were grown on both nutrient and nitrogen deficient media and supplied with prey have been shown to grow better than unfed plants (Behre 1929, Oudman 1936), while cytochemical and radio-tracer studies have demonstrated enzyme production and the transfer of products by the sessile glands of *Pinguicula* (Heslop-Harrison & Knox 1971).

Therefore insectivory does seem to benefit the nitrogen economy of these plants.

The symbiotic fixation of atmospheric nitrogen is a well-known characteristic of leguminous plants which has long been exploited by agriculturalists. Leguminous plants such as species of *Genista*, *Sarothamnus*, *Cytisus* and *Ulex* exist on poor, dry soils of little agricultural worth and although their presence must influence the nitrogen economy of such sites the effects have not been much investigated by ecologists. Nonleguminous plants which can fix atmospheric nitrogen also exist on nitrogen-deficient soils. *Alnus glutinosa* and *Myrica gale* have symbiotic associations which enable them to fix nitrogen in wet soils (Bond 1967) and many blue-green algae can fix atmospheric nitrogen either by themselves or in association with other organisms, as in some lichens (Stewart 1974).

In sand dunes, which are recent deposits and low in nitrogen, plants such as the nonleguminous sea-buckthorn have a symbiotic association which results in increased plant and soil nitrogen in stands of increasing age (Stewart & Pearson 1967). Nonleguminous nitrogen fixation may be important in other recently deposited soils. Species of *Casuarina* may colonize tropical dune systems whilst *Dryas drummondii* is an important colonizer of the nitrogen-poor debris left by retreating glaciers (Lawrence *et al.* 1967).

Various mycorrhizal associations, particularly those associated with ericaceous plants, have been credited with the ability to utilize atmospheric nitrogen although this now seems unlikely (Bond & Scott 1955, Harley 1969). However, mycorrhizas may enhance the abilities of plants to make use of alternative sources of nitrogen; for example, beech mycorrhizas can take up ammonium ions from dilute solutions although they have almost no capacity to absorb nitrate (Carrodus 1966).

Species also show adaptation to nitrogen supply by being able to use different sources of nitrogen. In acid soils nitrogen is available as ammonium ions whereas more nitrate is available in less acid soils. Species or ecotypes adapted to acid conditions show a greater ability to utilize ammonium than those from less acid sites although most plants show better growth when supplied with nitrate (Figure 7.6).

Species which are able to adapt to soils low in nitrogen will be able to exploit such soils at the expense of less well adapted species.

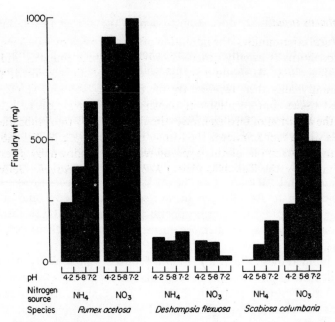

FIGURE 7.6. Dry matter production (mg) of different species with respect to nitrogen source (modified from Gigon & Rorison 1972).

The most acidophilous species (*Deschampsia flexuosa*) grows better with ammonium and is inhibited by nitrate at high pH whereas the other two species grow better with nitrate (particularly in less acid regimes) and are most inhibited by ammonium at low pH.

Conversely, some species will be limited to areas with a better nitrogen supply because they are unable to persist where supply is poor. Salt marsh plants such as *Salicornia* and *Sueda* benefit from added nitrogen and phosphorus, but nitrogen appears to be the limiting nutrient (Pigott 1969). In sand dunes the poor performance of *Ammophila arenaria* in dune pasture is associated with a lack of nitrogen; its better performance in areas where sand is being deposited is associated with improved root growth rather than with improved nitrogen supply (Willis 1965, cf. Bradshaw 1969). On the other hand, a species such as the nettle (*Urtica dioica*) that is traditionally associated with nitrogen-enriched sites is actually found to be more dependent upon soil phosphorus (Pigott & Taylor, 1964). Phosphorus supply has been extensively studied and is considered next.

Phosphorus supply

In natural communities, the limitation of plant growth by a deficiency of phosphorus is usually observed only when the element is added as fertilizer (e.g. Atkinson 1973). The characteristic species of the community may then respond by increased growth and show increased tissue concentrations of phosphorus (Figure 7.7): in other cases the addition of fertilizer may stimulate the growth of subsidiary species which then eliminate the former dominants (e.g. Willis 1963). A dramatic example of this type of response is shown by swards dominated by the Teesdale rarity, *Kobresia simpliciuscula* (Jeffrey

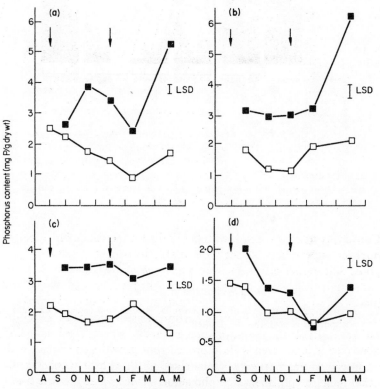

FIGURE 7.7. The total phosphorus levels in the leaves of four of the dune species, from August 1968 to May 1969: (a) *Carex panicea*; (b) *C. arenaria*; (c) *Hieracium pilosella*; (d) *Calluna vulgaris*. ■, Fertilized plants; □, control plants. The arrows indicate the times of application of fertilizer (Atkinson 1973).

& Pigott 1973). After the addition of phosphatic fertilizer, the grasses *Festuca rubra*, *F. ovina* and *Agrostis stolonifera* increase in amount with a corresponding decrease in the representation of *Kobresia*. In this case the rare species can survive only when deficiency limits the competitive ability of other species.

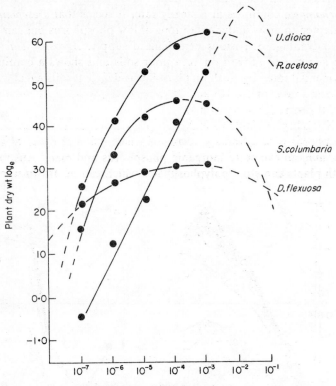

FIGURE 7.8. Responses of four dissimilar species to a range of external phosphorus concentrations. Dotted lines suggest trends beyond the range of concentrations used (Rorison 1969).

Species differ markedly in their responses to phosphorus concentration (Figure 7.8). Increased phosphorus results in increased growth by nettle (*Urtica dioica*) over a wide range of phosphorus concentrations whilst the grass, *Deschampsia flexuosa*, shows little increase. Other species (e.g. *Rumex acetosa*, *Scabiosa columbaria*) show a decreased response at high concentrations of phosphorus. *Urtica* is

characteristic of nutrient-rich sites whereas *Deschampsia* is typical
of poor soils. Isolated roots of the two species take up similar
amounts of phosphorus (Nassery & Harley, 1969)—this is utilized
in shoot growth in *Urtica* but remains as inorganic phosphorus in
the shoots of *Deschampsia*. Consequently, the growth of *Urtica* is
restricted when the plants are deprived of phosphorus but that of
Deschampsia continues at a steady rate. It seems that *Deschampsia*
is adapted to grow under conditions of low phosphorus but can also
take advantage of periods of enhanced supply. *Agrostis setacea* is
also adapted to growth on very poor soils and shows a continued
exponential rate of growth on phosphorus-deficient soils which cause
decreased rates of growth and deficiency symptoms in *A. stolonifera*
and *A. canina* (Clarkson 1967).

The ability of plants to make use of or store transient supplies of
phosphorus is of advantage to plants which colonize deficient soils.
Deschampsia can store inorganic phosphorus and many Australian
heath plants can store polyphosphates which can be drawn upon in

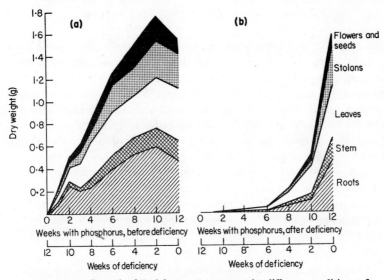

FIGURE 7.9. Growth of *Epilobium montanum* under different conditions of
phosphorus supply. (Atkinson & Davison 1971).
  (a) Periods of 0–12 weeks with a phosphorus supply followed by 12–0
      weeks with no supply.
  (b) Periods of 0–12 weeks of phosphorus deficiency followed by 12–0
      weeks of supply.

times of shortage (Jeffrey 1964, 1968). Mycorrhizas may provide a similar strategic reserve of phosphorus (Harley 1969, Bannister & Norton 1974). The supply of phosphorus may be more influential when it is experienced early in the growth of the plant. In the weed, *Epilobium montanum*, an initial supply of phosphorus for only one week enabled the plant ultimately to produce stolons whilst two weeks' supply allowed seed production. Much more phosphorus was needed when plants were initially starved of the element. Plants deprived for the first six weeks required a further six weeks of supply in order to produce stolons, whereas plants starved for only four weeks needed eight weeks' supply in order to produce seed (Figure 7.9). An adaptation such as this would enable a weed to exploit a local area of supply in an otherwise deficient soil. Local enhancement of supply could be the result of microbial distribution, the death of individual plants and animals, the deposition of urine or faeces and even the dropping of scraps of food and cigarette ends.

7.4 NUTRIENT EXCESS AND MINERAL TOXICITY

An excess of an element may cause a reduction in the growth of a plant species. The element in question may be essential for growth in smaller amounts (e.g. magnesium, copper, zinc) but can be non-essential (e.g. lead, nickel, chromium). Such elements are often termed 'toxic' when they inhibit growth although strictly this term should be used only when the inhibition is specifically due to a particular element. In many cases the excess of one chemical can lead to the deficiency of others as in calcareous soils where the super-abundance of calcium carbonate may be associated with deficiencies of phosphorus and iron. The measurement of toxicities by growing whole plants is time-consuming and rapid techniques may involve the short-term effects of solutions upon root growth (Gregory & Bradshaw 1965, Sparling 1967) or an examination of the effects of ions on the properties of cells (Repp 1963).

Excesses of particular chemicals may have a geological or geographical origin as in the excess of calcium carbonate in soils derived from limestones or of salt in soils influenced by the sea. The origin may be biotic as in the guano of sea-bird colonies or anthropogenic as in sewage production, mining and industrial activity. Long-standing habitats often have characteristic assemblages of species which have evolved in response to the particular habitat. Thus there are species which are described as calcicolous (lime-dwelling) or halophilous (salt-loving): there are also distinct intraspecific races (ecotypes) which are adapted to specific soils. These can occur on soils of more recent origin as well as on older sites. The edaphic ecotype from an extreme soil may show its best growth on 'normal' soil but be able to tolerate the extreme conditions to a greater degree than less well-adapted ecotypes (Figure 7.10). Plants adapted to excessive amounts of a particular element may be merely tolerant of the excess, but they could also have a requirement for a large amount of the element or be adapted to any deficiencies of other elements that are induced by the original excess. The adaptation to excess may also induce characteristics which exclude adapted species from normal soils. These considerations will be illustrated by considering adaptation to calcareous soils, saline environments and excesses of heavy metals.

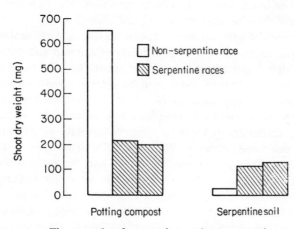

FIGURE 7.10. The growth of serpentine and nonserpentine races of *Rumex acetosa* on serpentine and 'normal' soil (data of Proctor 1971a).

7.4.1 *Calcicoles and calcifuges*

There is a strong relationship between the amount of calcium in a soil and its pH (Figure 7.11). This has led to a confusion between the separate effects of calcium and pH and a tendency to term all plants from soils with a high pH 'calcicolous' (i.e. lime-dwelling) and those from soils with low pH as 'calcifugous' (i.e. lime-fleeing). Some

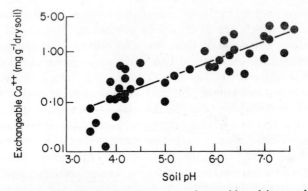

FIGURE 7.11. The relationship between exchangeable calcium and pH in a range of heathland soils from Northern Spain (unpublished data).

ecologists have attempted to avoid this confusion by using the term 'basiphilous' (base-loving) for plants which occur on soils of high pH but low in calcium (e.g. serpentinic soils). The opposite of this term would then be 'acidophilous' (acid-loving).

If the effects of calcium alone are considered, then calcicoles would be expected to be tolerant of high levels of calcium and possibly to have a high requirement for the element. Calcifuges would be expected to have low tolerance and be able to subsist on soils low in calcium. Calcifuges such as *Juncus squarrosus* and *Nardus stricta* show their best growth in solutions low in calcium whereas the calcicole, marjoram (*Origanum vulgare*) appears to require high levels of calcium (Jeffries & Willis 1964). A similar contrast is seen between the calcifugous *Agrostis setacea* and the more calcicolous *A. stolonifera* (Figure 7.12). However, *Nardus stricta* is capable of growing in dilute solutions with a high ratio of calcium to other ions whereas the growth of even *Origanum* is restricted on concentrated solutions with a similar ratio of calcium to other ions (Jeffries & Willis 1964).

The effects of calcium may be independent of pH. *Agrostis tenuis*, *A. stolonifera* and *A. canina* all show an increase in root growth when calcium sulphate is added to a deficient soil (without a change in pH). The addition of calcium carbonate results in increased growth of both roots and shoots, increased tissue concentrations of calcium in all three species and a rise in soil pH from 4·1 to 5·9. In contrast *A. setacea* shows no increase in either growth or tissue concentrations when calcium is added (Clarkson 1965).

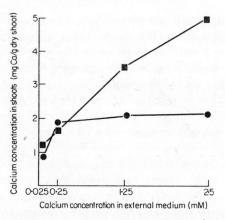

FIGURE 7.12. Calcium concentration in the shoots of *Agrostis* species as a function of calcium in the external medium. ●, *A. setacea*; ■, *A. stolonifera*. (Clarkson 1965).

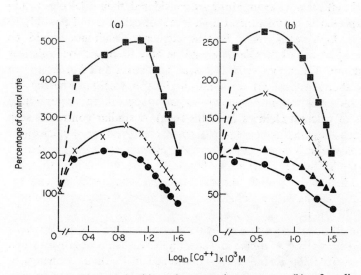

FIGURE 7.13. Phenotypic (a) and genotypic responses (b) of malic dehydrogenase from clones of *Lemna minor*. (Jefferies *et al.* 1969).
  (a) Response of enzymes from plants growing on solutions with Ca^{++} concentrations of $10^{-2}M$ (■), $10^{-3}M$ (×) and $10^{-4}M$ (●).
  (b) Response of clones from different habitats. Acidic races (▲, ●), calcareous races (×, ■).

Evidence for direct calcium toxicity may be confused with effects due to the deficiency of iron (Grime & Hodgson 1969) but the rapid death of seedlings is unlikely to be caused by deficiency alone and calcium could be inhibitory to enzyme systems at high concentrations as it would compete with other ions (e.g. magnesium and potassium) for cofactor sites on particular enzymes (Jeffries & Willis 1964). This is borne out by studies of the activity of malic dehydrogenase with regard to calcium concentration. The enzyme is more inhibited by high concentrations in extracts from acidic races of *Lemna minor* and from clones which have grown on solutions low in calcium (Figure 7.13). Plants may avoid high concentrations of calcium in cell sap by binding it as oxalate, as in many Caryophyllaceae, or may tolerate large amounts of free calcium as in many Leguminosae. Other species characteristic of calcareous soils develop a high ratio of potassium to calcium (Horak & Kinzel 1971) and presumably exclude calcium.

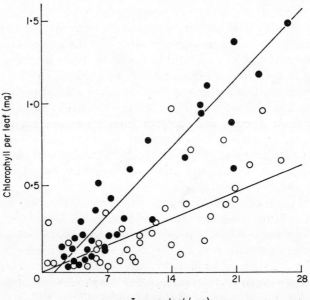

FIGURE 7.14. The relationship between leaf chlorophyll and leaf iron in a calcareous (●) and a noncalcareous (○) race of *Teucrium scorodonia* grown on a calcareous soil (Hutchinson 1967).

Calcium toxicity may be responsible for excluding some species from calcareous soils. A characteristic lime-chlorosis is induced in most species from acidic habitats but rarely in those from calcareous soils (Grime & Hodgson 1969). Plants which are susceptible to lime chlorosis are also sensitive to iron deficiency. The ratio of chlorophyll to iron content is much lower in plants susceptible to lime chlorosis than in resistant plants when both are grown on iron-deficient solutions (Figure 7.14). The bicarbonate ion interferes with the uptake of iron and induces chlorosis even in strains of *Teucrium scorodonia* which are otherwise resistant to lime chlorosis and the resistant ecotype may depend upon iron accumulated in its roots (Hutchinson 1968). Other species may differ in their susceptibility to the bicarbonate ion (Lee & Woolhouse 1969) and bicarbonate toxicity may be another factor in excluding calcifuges from calcareous soils. Chlorotic and green individuals may not show large differences in their overall growth in controlled conditions but chlorotic individuals show a poor control of water loss which may render them more susceptible to drought in the field (Hutchinson 1970a,b).

The exclusion of calcicoles from acid soils may be a straight case of calcium deficiency, although in many instances calcicoles are capable of growth on media low in calcium if the concentration of all other nutrients is low (Jeffries & Willis 1964). Currently the presence of toxic ions in acid solutions is considered to be a major factor. Ions of metals such as copper, manganese, iron and aluminium all occur and of these aluminium ions appear to be the most universally toxic. Species and ecotypes from calcareous habitats are usually more sensitive to aluminium toxicity than those from acidic habitats and there is a general correlation between the occurrence of a species on acidic sites and its tolerance of aluminium (Figure 7.15).

The absence of aluminium ions has been used to explain the presence of the otherwise basiphilous rush, *Schoenus nigricans* in the acid bogs of Western Ireland (Sparling 1967). *Schoenus* is sensitive to aluminium toxicity, but the ion, which is mostly derived from dust, is less prevalent in the oceanic climate of Western Ireland. However, Boatman (1972) considers that it is the absence of grazing rather than the absence of aluminium that allows the species to persist on blanket peat in Ireland.

In conclusion, it seems that various combinations of a high requirement for calcium, a tolerance of calcium and bicarbonate

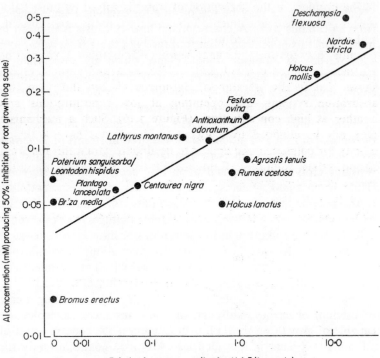

FIGURE 7.15. The relationship between the aluminium tolerance of various species and their relative occurrence of soils of < pH 4·5. (Adapted from Grime & Hodgson 1969).

(The relative frequency is the ratio of the occurrence of a species on soils of < pH 4·5 to that on soils of > pH 4·5. A constant factor of 0·01 has been added to all values so that zero can be represented on the logarithmic scale).

toxicity, a low requirement for iron and a sensitivity to toxic ions such as aluminium characterize calcicoles. Calcifuges tend to show the opposite characteristics.

7.4.2 *Halophily and salt tolerance*

Halophytes are obliged to cope with excessive amounts of salts (particularly sodium chloride) in their environment. This excess presents difficulties in the acquisition of other mineral nutrients and water and the accumulation of ions within tissues may also be toxic.

The problem of the absorbtion of water is solved by most halophytes by osmotic adjustment. Sufficient ions, particularly chloride
and sulphate are absorbed in order to maintain a gradient of water
potential between the plant and the soil (Walter 1960). Ions must be
abstracted from a concentrated soil solution and Epstein (1969)
shows that in the wheatgrass, *Agropyron elatium* there are two
absorbtion systems—one operating at low concentrations and
another at high concentrations (Figure 7.16). Such a mechanism
may not be confined to halophytes. Halophytes have a strong
affinity for potassium and appear to be able to extract this ion from
strong saline solutions (Rains & Epstein 1967). On nonsaline
media potassium may substitute for sodium and be absorbed in
large amounts (e.g. in *Aster tripolium* and *Plantago maritima*) but
other species (e.g. *Lepidium crassifolium*) appear to be obligate

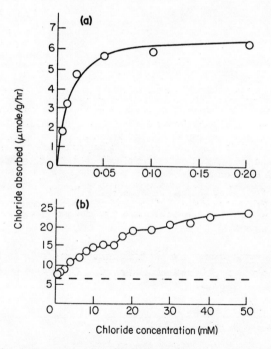

FIGURE 7.16. Rate of uptake of chloride, at low (a) and high concentrations of KC1, by tall wheatgrass, *Agropyron elongatum* (Epstein 1969).
    Note the differences in scale on the two graphs. The dashed line in (b)
indicates the maximal absorbtion by the mechanism in (a).

halotypes and are unable to grow well on soils deficient in sodium (Weissenböck 1969).

The proportions of ions absorbed by halophytes are not constant —they vary on different soils (Weissenböck 1969) and with season (Capeletti Paganelli 1968). The ratio of amounts of sodium, potassium and calcium to the amount of chlorine has been used as an index of halophytism (halophytes normally have a value below 2·0— (Figure 7.17). The ratio varies with season: species such as *Limonium vulgare* show a minimum ratio in midsummer whereas *Salicornia herbacea* and *Convolvulus soldanella* show minima at the beginning and end of the growing season respectively (Capeletti Paganeli 1968). In general the more halophytic species show proportionately smaller annual fluctuations in their ionic relations than nonhalophytes (Figure 7.17).

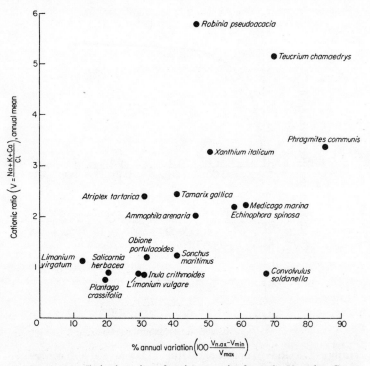

FIGURE 7.17. Cationic ratios of various species from the Venetian Coast as a comparative index of halophytism (data of Capaletti Paganelli 1968).

Many plants possess mechanisms of salt avoidance. Salt glands secrete salt from the leaves of many genera (e.g. *Spartina, Glaux, Armeria, Limonium*). Salt secretion is not a universal adaptation to a particular saline environment. Some mangrove genera (e.g. *Avicennia, Aegiceras*) secrete salt whilst others (e.g. *Rhizophora*) do not. The process of salt secretion suggests that active (i.e. requiring metabolic energy) transfer of salt or active absorption of water occurs (Scholander *et al.* 1962).

Succulence represents another method of salt avoidance as the expansion of leaf volume allows a continued uptake of salts without large increases in concentration (Walter 1960). The chloride ion is usually considered to be important in inducing succulence (van Eijk 1939) although the substitution of potassium for sodium is associated with a decrease in succulence (Weissenböck 1969). Some authors (e.g. Jennings 1968) consider sodium to be important in determining succulence. Some plants (e.g. *Juncus gerardii, J. maritimus*) get rid of excess salt by the wholesale shedding of leaves and some desert species of *Atriplex* shed and replace epidermal hairs which accumulate chloride.

The damage that occurs when plants are treated with salt is largely a function of the chloride ion—salts such as sodium and calcium chloride have comparable effects (Buchsbom 1968). Species are generally most resistant to salt in the winter and have a low resistance in summer with a minimum at the beginning of the growing season. In this respect salt resistance is similar to other tolerances (cf. drought, heat and cold resistance). In winter some nonhalophytes (glycophytes) are equally or more resistant to salt than many halophytes. However, halophytes are usually more resistant; for example, the halophytic shrubs, *Tamarix tetrandra* and *Halimodendron halodendron* are more resistant than the glycophytic *Cornus alba* and *Prunus avium*. There are large differences in the salt tolerance of related species. *Acer platanoides* and *A. pseudoplatanus* are highly resistant whereas *A. campestre* is less resistant and *A. negundo* is less resistant still. A knowledge of salt tolerance can have an applied relevance, particularly for the roadside planting of trees in regions where salt is applied to clear ice and snow. Deciduous trees are generally more resistant than evergreens and conifers. Buchsbom (1969) has examined the salt tolerance of a great number of woody species (Table 7.1).

The application of salt to vegetation generally reduces the growth

TABLE 7.1. Salt tolerance in various woody species in winter (abstracted from Buchsbom 1968).

Tolerant	Sensitive	Very sensitive
Acer platanoides	*Acer negundo*	*Chamaecyparis pisifera*
A. pseudoplatanus	*Betula pendula*	*Cornus alba*
Aesculus hippocastanum	*Hippophae rhamnoides*	*Crataegus monogyna*
Alnus glutinosa	*Ligustrum vulgare*	*Fagus sylvatica*
Fraxinus excelsior	*Prunus avium*	*Hedera helix*
Larix decidua	*P. padus*	*Ilex aquifolium*
L. leptolepis	*Populus tremula*	*Picea abies*
Malus sylvestris	*Salix aurita*	*Pinus sylvestris*
Ribes nigrum	*S. caprea*	*Platanus acerifolia*
Robinia pseudoacacia	*Sorbus aria*	*Quercus robur*
Tilia cordata	*Tsuga heterophylla*	*Q. petraea*
		Taxus baccata

of species and may alter the outcome of competition. When common roadside grasses and clover are grown in mixtures, the application of salt has a lesser effect upon the performance of *Cynosurus cristatus*, *Festuca rubra*, and *Lolium perenne* than upon *Trifolium repens* and *Poa pratensis* (Mitchell 1973). The competitive balance between plants that are normally subjected to salt may also be influenced. *Festuca rubra* from coastal cliffs becomes dominant over *Armeria maritima* in mixtures watered with tapwater, whereas *Armeria* becomes dominant (despite reduced growth of both species) when mixtures are watered with saltwater (Goldsmith 1973). This relationship explains the development of *Armeria* swards on sites more subject to salt spray.

7.4.3   *Excess of metals*

The fact that plants occur at all on soils with high or abnormal concentrations of metallic ions indicates that they must be adapted to the peculiar conditions of the site. Plants may avoid any toxicities by colonizing areas of lower metal concentration within the contaminated area or avoid, through some exclusion mechanism or selective uptake, excessive cellular concentrations of metallic ions (Antonovics *et al.* 1971). However, in many instances, high concentrations of metals are found in the tissues of plants growing on contaminated soils (Table 7.2) and such plants are considered to be metal tolerant.

TABLE 7.2. Heavy metal concentrations in the above-ground portions of various species growing on contaminated soils (maximal values in ppm dry tissue).

Species	metal					
	Zn	Pb	Ni	Cr	Cu	Co
Minuartia verna	3 200	1 580	220	150	338	106
Cerastium holosteoides	960	2 200	190	50	—	27
Thlaspi alpestre	8 400	640	—	—	90	—
Cochlearia officinalis	—	—	420	145	—	58
Agrostis tenuis	3 500	1 950	—	—	—	—
A. canina	110	570	187	88	—	134
Calluna vulgaris	170*	2 200	190	5	120*	27

Abstracted from Ernst (1965, 1968)—lead/zinc/(copper) soil and Johnston (1974)—lead/zinc and serpentine soil.

*R. Marrs, unpublished data—lead/zinc and copper spoil.

Concentrations in roots may be much greater and may prevent excessive concentrations in above-ground portions. For example, Wu & Antonovics (1975) found maximal levels of copper of about 21 ppm in shoots but 3,275 ppm in roots of *Agrostis stolonifera*; corresponding values for zinc were 64 and 14 750 ppm respectively.

In Europe, certain species such as *Viola calaminaria*, *Thlaspi alpestre*, *Minuartia verna*, and *Armeria* spp. are very characteristic of metalliferous sites (Ernst 1965a). Other species have been used as indicators of metallic ores elsewhere (Antonovics *et al.* 1971). The existence of such species suggests that speciation has occurred on soils with abnormal metal contents. This is borne out by the examination of species found on more recently deposited mine spoil where tolerant ecotypes of more widespread species (e.g. *Silene vulgaris*, and grasses such as *Festuca ovina*, *Agrostis spp.*, and *Anthoxanthum odoratum*) occur. The degree of tolerance exhibited by such ecotypes is often correlated with the amount of metal found in the soil (Figure 7.18). Tolerance is genetically determined and appears to be a dominant, polygenic, characteristic (Gartside & McNeilly 1974a,b). Normal populations of plants which develop tolerant races contain a low frequency (< 2 per cent) of tolerant individuals which are selected out on contaminated soils (Walley *et al.* 1974, Gartside & McNeilly 1974c). Metal tolerant species may also be tolerant of nutrient deficiencies. In *Silene vulgaris* increased tissue concentrations of zinc are associated with decreased phosphorus concentrations (Ernst

1968b) and the presence of lead appears to decrease the availability of phosphorus on contaminated soils (Jeffrey 1971).

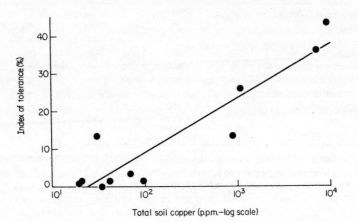

FIGURE 7.18. Relationship between copper tolerance and soil copper for populations of *Mimulus guttatus* (data of Allen & Sheppard 1971).

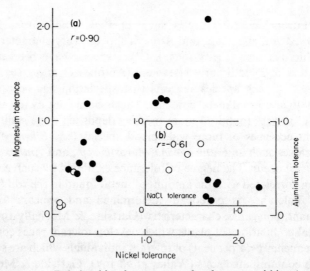

FIGURE 7.19. Relationships between metal tolerances within *Agrostis stolonifera* (a) and *Festuca rubra* (b). (Data of Proctor 1971b and Hunter 1971).
Open circles relate to nonserpentine and nonsaline sites respectively.

Metal tolerance is very specific. Plants which are tolerant of a particular ion are rarely found to be tolerant (co-tolerant) of ions of similar physico-chemical characteristics which are absent from the soils on which they grow (Turner 1969). Multiple tolerances occur when a number of ions are present in the same soil as in clones of *Agrostis stolonifera* which are tolerant of both magnesium and nickel (Figure 7.19a). Nickel and magnesium have similar ionic sizes and thus co-tolerance is not impossible, but all sites with a high magnesium content do not necessarily support races with a high nickel tolerance and vice versa (Proctor 1971b). The tolerance mechanisms for each element appear to be specific and separate, even in plants showing multiple tolerance (Wu & Antonovics 1975).

A large number of metal-tolerant races have been described (Antonovics *et al.* 1971). The adaptability within a genus is illustrated by the various tolerant races that have been recorded for *Agrostis* spp. (Table 7.3). Most of the tolerances are for the so-called heavy metals (defined as those with a density of greater than 5g cm^{-3}) but tolerance of lighter metals such as magnesium and aluminium also exists.

TABLE 7.3. Metal tolerances in *Agrostis* spp.

species	Mg	Al	metal Cr	Ni	Cu	Zn	Pb
Agrostis tenuis	.	(+)	.	±	±	±	±
A. canina	±	(+)	±	±	.	.	±
A. stolonifera	±	−	±	±	±	±	.
A. setacea	.	+	.	.	.	.	.

Data of Clarkson 1966, Turner 1969, Proctor 1971b, Craig 1972 and Wu & Antonovics 1975. Dots indicate lack of information.

+ tolerant, (+) somewhat tolerant, − intolerant, ± indicates that both tolerant and non-tolerant races have been demonstrated.

Races of *A. canina* and *A. stolonifera* from serpentine soils show multiple tolerance of Mg, Cr and Ni (Proctor 1971b) and the race of *A. stolonifera* used by Wu & Antonovics was tolerant of both copper and zinc; otherwise races tolerant of a specific metal have been demonstrated. The data for Al (Clarkson 1966) is based on interspecific comparisons only.

Metal tolerant races are usually excluded from normal soils by competition from the more vigorous nontolerant ecotypes (e.g. Antonovics 1966). In many instances both tolerant and nontolerant plants grow better on normal soils with the best growth being shown by the nontolerant plants, whereas the reduction in growth on contaminated soils is less in the tolerant varieties giving them a competitive advantage (Figure 7.10). It is also possible that tolerant ecotypes have a greater requirement for metals than nontolerant types. In some instances the growth of roots of tolerant types is stimulated in solutions containing metals (Grime & Hodgson 1969, Antonovics et al. 1971, Proctor 1971b). The metallic ions possibly substitute for trace elements, but any requirement for a large amount of metallic ions could exclude a tolerant plant from normal soils. Tolerant plants may also be excluded from other soils because they are intolerant of ions found there. Clones of *Festuca rubra* from a salt marsh have a high salt tolerance but are sensitive to aluminium whereas the converse is true of clones from acid pasture (Figure 7.19b).

7.5   PLANT RESPONSES TO 'NORMAL' NUTRIENT CONDITIONS

It should by now be apparent that it is impossible to define a 'normal' nutrient regime for plants. A regime which may promote optimum growth of one species or genotype may cause reduced growth as a result of toxicity or deficiency in another. Most soils support vegetation composed of species that are adapted to the prevailing conditions and which are tolerant of any excesses or deficiencies that may occur. Ecotypic differentiation can take place over a relatively short period of time. Davies & Snaydon (1973a,b, 1974) have examined the responses of *Anthoxanthum odoratum* from the Park Grass Experiment (Thurston 1969) and found that populations from differently treated plots differ in their responses to calcium, aluminium and phosphorus. There is a graded response which is usually related to the availability of the particular element in the soil and modified by factors such as soil pH and liming. The selection and evolution of these races has taken place in only 65 years.

The precise mixture of species that grows on any normal soil will be decided by competition. When ecotypes of different species are equally tolerant of the mineral status of the site, then competition for other factors such as light and water have an overriding importance (van den Bergh 1969). Usually the more vigorous competitor wins out in a competitive situation, although in some cases the lower yielding competitor may win. Calcareous ecotypes of *Trifolium repens* grow as well or better than acidic races on both calcareous and acidic soils both singly and in competition, but the acidic race performs better than the calcareous race when implanted into natural grassland, even on limed plots (Snaydon 1962). The effects of competition are considered in more detail in the next chapter.

8: Interactions

8.1.1   *Competition for space, light and nutrients*

There is often a marked difference between physiological and ecological responses than can frequently be related to competition or 'interference' by other plants (e.g. Rorison 1969) and consequently the experimental study of interspecific competition is relevant to physiological studies.

The standard method of examining the response of competing species is the replacement series of de Wit (1960) in which two (or more) species are grown together at the same overall density but in varying proportions. A simple replacement series (e.g. Shontz & Shontz 1972) with two species (A,B) at overall density (m) would have only three treatments

$$\text{i.e. } m(\text{A}), \tfrac{1}{2}m(\text{A}) + \tfrac{1}{2}m(\text{B}), m(\text{B}).$$

The mathematical, as opposed to statistical treatment, of these results is dealt with by de Wit (1960) and extended by de Wit & van den Bergh (1965). The full statistical analysis of such data is more difficult (e.g. McGilchrist & Trenbath 1971) and is not always attempted.

There are a number of possible outcomes for a replacement series (Figure 8.1). There may be no effect or one species may suppress the other. It is also possible for growth in mixtures to cause mutual stimulation (e.g. Bakhuis & Kleter 1965) or depression (van den Bergh & Elberse 1962). As the growth of individuals of a species is usually depressed by other members of the same species (intraspecific competition), the outcome of competition between two species may be interpreted in terms of the balance between intra- and interspecific competition. Individuals of species in which intraspecific competition is dominant may show better growth in mixed cultures where intraspecific effects are lessened; whereas species more

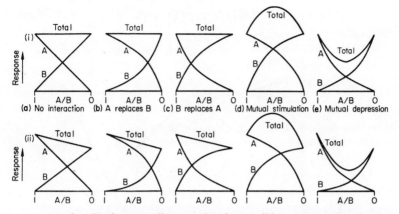

FIGURE 8.1. Replacement diagrams showing possible outcomes when two species are grown together. In series (i) the yield of both species in monoculture is identical, in (ii) species B is lower yielding and thus ii (c) gives an example of the 'Montgomery Effect'. The vertical axis is some measure of species' response (e.g. dry weight) and the horizontal scale is of the initial relative proportions of the two species.

sensitive to interspecific competition will show their best growth in pure cultures. The interpretation of a replacement series becomes further complicated if one species grows less well in pure culture than the other but even so it is possible for the lower-yielding species to suppress the higher-yielding species (e.g. van den Bergh 1969). This is the so-called (Montgomery Effect) (Montgomery 1912).

The outcome of competition between two species depends upon the prevailing environmental conditions and the competitive balance may be modified in a changed environment. For example, *Festuca rubra* suppresses *Armeria maritima* in mixtures watered with tap water whereas, although growth of both species was much reduced, *Armeria* suppresses *Festuca* in mixtures watered with sea water (Figure 8.2). Low-yielding species may have a competitive advantage under extreme environmental conditions. Thus, the grasses *Alopecurus pratensis*, *Anthoxanthum odoratum* and *Agrostis tenuis* replace the higher-yielding *Lolium perenne* and *Dactylis glomerata* on infertile soils although the converse is true on fertile soils (van den Bergh 1969). Metal tolerant ecotypes are often competitively superior on metalliferous soils and soils low in phosphorus but inferior on normal soils (Antonovics *et al.* 1971). Such changes in response have been interpreted as a shift from competition for space above ground in favourable conditions to root competition in less favourable soils.

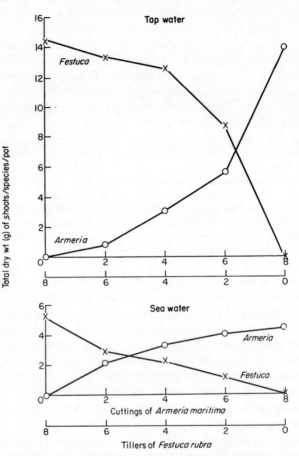

FIGURE 8.2. Results of a competition experiment between *Festuca rubra* and *Armeria maritima* when watered with tap water and sea water. Each point is the mean of two replicates. (After Goldsmith 1973).

The complexity of interpreting the mechanisms of competition can be illustrated by the phenomenon of 'heather check' in spruce plantations where the growth of spruce and birch trees with an understory of heather is often severely retarded. The growth of the trees is much improved when inorganic fertilizers (particularly those containing nitrogen) are applied or if the heather is removed. This suggests that the heather is competing with the spruce for nutrients. However, Handley (1963) showed that extracts from heather litter inhibited spruce mycorrhizal fungi and Robinson (1972) considered that the fungitoxic principle was derived from living mycorrhizal heather roots. Extracts of heather litter have also been shown to be phytotoxic (Robinson 1971). Thus, the suppression of spruce can be explained as direct competition between the two species for mineral nutrients, competition between their mycorrhizas, or as chemical inhibition of either tree growth or mycorrhizal growth. Only critical experimentation could differentiate between these various possibilities.

8.1.2 Chemical antagonism

Extracts of living or dead plant material often can be shown to inhibit the growth of other plants. Molisch (1937) coined the word allelopathy for this phenomenon. Simple water-soluble substances often appear to be effective as in the case of humus extracts from under stands of dog's mercury (*Mercurialis perennis*) which inhibit the development of test seedlings (Figure 8.3) and extracts of living and dead tissue of *Allium ursinum* which also inhibit the growth of test plants (Lange & Kanzow, 1965). Both *Allium* and *Mercurialis* form dense stands in woodland with very few associated species and it could be assumed from the data that chemical inhibition has some role to play in eliminating competitors.

However, other factors could have overriding effects. The closed nature of stands might merely be a result of the plants' growth form, their method of vegetative reproduction and the exclusion of other species as a result of competition for light and nutrients. Chemical antagonism can only be convincingly demonstrated when inhibitory materials are shown to be produced, to be active at their normal concentrations and to remain effective in the soil. The effects of competition for light, nutrients, water and any other biotic effects must not account for the observed differences (Whittaker 1970). These criteria are not always fully met, and even if they are, ecologists seem reluctant to accept that chemical antagonism has a significant role. Microbiologists, on the other hand, recognize that chemical antagonisms occur between soil micro-organisms (e.g. Park 1958) and also that higher plants may produce substances, phytoalexins, which inhibit pathogens (Kuć 1972). Whittaker (1970) suggests that

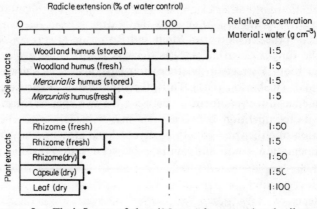

FIGURE 8.3. The influence of plant (*Mercurialis perennis*) and soil extracts on radicle extension in lettuce (data of Stevenson 1972).

Notice the inhibition caused by many extracts associated with Mercurialis. (*Significant difference from control in distilled water).

differences in attitude may stem from a consideration that green plants are essentially pure and harmless whereas micro-organisms are acknowledged to be noxious.

There is no *a priori* reason why plants should not produce harmful secondary or waste substances and polyphenols, terpenoids, alkaloids and nitriles have all been shown to exist. These substances may be released from above and below ground parts of the plant and from living or dead material and there is evidence to support such a variety of sources (Table 8.1).

TABLE 8.1. Some substances implicated in chemical antagonisms between higher plants.

Origin	Substance	Plant Organ	Species	Reference
Volatilization	terpenoid	leaves	sagebrush	Muller 1965, Muller & Hauge 1967
Leaching	phenolic (juglone)	leaves	walnut	Bode 1958
Decay	glycoside (amygdalin)	roots (bark)	peach	Borner 1960
Exudation	phenolic (scopoletin)	roots	cereals	Martin & Rademacher 1960

There are many other examples of allelopathy (Whittaker 1970a, Went 1970) but the question of their efficacy in the field remains. Whittaker (1970a) argues that the fact that many plants appear tolerant of secondary chemicals does not preclude their importance; most plants are adapted to the prevailing chemical and physical environment which ecologists have always considered to be important He has extended the role of secondary substances (1970b) and suggests that they may be essential regulators in the growth and development of ecosystems, analogous to hormones in the organism. Went (1970), however, considers that allelopathy is effective only in dry climates and the best documented evidence does come from areas of arid vegetation, in particular chaparral. The investigation of the effects of *Adenostoma fasiculatum* provides a good example (McPherson & Muller 1969). The addition of fertilizer and plant ash did not stimulate the germination and establishment of seedlings under *Adenostoma* bushes but the complete removal of the vegetative canopy did. Shrubless plots which were artificially shaded to the same degree as under the shrubs showed no significant reduction in seedling establishment and levels of soil moisture were similar underneath and outside the canopy. There was no difference in seedling establishment under the shrubs after a wet and dry season. Grazing by small animals did have a minor effect (cf. Bartholemew 1970) but large numbers of seedlings were produced only when the soil was heated although heat did not stimulate the germination of individual species outside the soil. The authors consequently concluded that an inhibitory substance was leached from the crowns into the soil and they then showed that natural and artificial leachates from the crowns were inhibitory to *Bromus rigidius*, a common local weed, and that this inhibition was retained in the soil.

This is a good example of what was essentially a field investigation. Other systems have been devised for laboratory investigations. Martin and Rademacher (1960) used a continuous solution culture whilst Jarvis (1965) made continuous soil leachates to compare soils with and without the suspect plant. Stevenson (1972) attempted to separate the effects of root competition from chemical antagonism by growing a test species within a cage which excluded the roots of the antagonizing species but allowed the transfer of any inhibitory substances. Substances which are leached from the leaves have been investigated by comparing the effects of top and bottom watering

(Grümmer & Beyer, 1960). Allelopathy will doubtless remain a controversial subject until sufficient critical work has been carried out. It is a field ripe for investigations.

8.2   INTERACTIONS WITH MICRO-ORGANISMS

8.2.1   *Rhizosphere and other effects*

Plants have a considerable influence on the presence of micro-organisms—they provide a surface for microbial growth and add organic matter and exudates to the soil environment in particular. There is a large stimulation of the bacterial population and in particular ammonifying and denitrifying bacteria (Table 8.2), other groups that are stimulated include groups capable of using amino acids, sugars and carbohydrates. Fungi and actinomycetes may also be stimulated in the vicinity of plant roots but other groups of soil organisms such as protozoa and algae may show little difference. This stimulation of microbial growth in the region of the roots (i.e. the

TABLE 8.2. The rhizosphere effect in spring wheat (Rouatt *et al.* 1960)

Microbial group	Numbers in rhizosphere soil (R)	Numbers in control soils (S)	Approximate ratio R:S
Algae	$5 \cdot 0 \times 10^3$	$2 \cdot 7 \times 10^4$	$1:5$
Protozoa	$2 \cdot 4 \times 10^3$	$1 \cdot 0 \times 10^3$	$2:1$
Fungi	$1 \cdot 2 \times 10^6$	$1 \cdot 0 \times 10^5$	$12:1$
Actinomycetes	$4 \cdot 6 \times 10^7$	$7 \cdot 0 \times 10^6$	$7:1$
Bacteria (total)	$1 \cdot 2 \times 10^9$	$5 \cdot 3 \times 10^7$	$23:1$
Bacterial groups			
Ammonifiers	$5 \cdot 0 \times 10^8$	$4 \cdot 0 \times 10^6$	$125:1$
Gas-producing anaerobes	$3 \cdot 9 \times 10^5$	$3 \cdot 0 \times 10^4$	$13:1$
Anaerobes	$1 \cdot 2 \times 10^7$	$6 \cdot 0 \times 10^6$	$2:1$
Denitrifiers	$1 \cdot 3 \times 10^8$	$1 \cdot 0 \times 10^5$	$1\ 260:1$
Cellulose aerobic	$7 \cdot 0 \times 10^5$	$1 \cdot 0 \times 10^5$	$7:1$
Decomposers anaerobic	$9 \cdot 0 \times 10^3$	$3 \cdot 0 \times 10^3$	$1:1*$
Spore-formers	$9 \cdot 3 \times 10^5$	$5 \cdot 8 \times 10^5$	$1:1*$
'Radiobacter' types	$1 \cdot 7 \times 10^7$	$1 \cdot 0 \times 10^4$	$1\ 700:1$
Azotobacter	$< 10^3$	$< 10^3$	?

*These ratios are listed as $1:1$ because the figures for rhizosphere and control soils show no statistically significant differences.

rhizosphere) results in increased respiration rates for such soils. The rhizosphere effect is more marked in light soils than in clays or organic soils and large increases in microflora are seen in the vicinity of desert and sand-dune plants (Webley et al. 1952).

The rhizosphere has a considerable effect on the mineral nutrition of plants. Ammonification of organic nitrogen represents a source of nitrogen which may be available to the plant, especially through subsequent nitrification, whereas denitrification will result in a loss of nitrate. Denitrification may be particularly acute in waterlogged conditions where nitrate may substitute for oxygen as a hydrogen acceptor. Phosphates may be immobilized by micro-organisms at low external concentrations (Barber & Loughman 1967) and at high pH (Barber 1969) although other micro-organisms may aid the dissolution of phosphorus (Louw & Webley 1959). The different balances set up between increased availability and immobilization may account for the different results of workers in this field (e.g. Rovira & Bowen 1966). The microflora may also be responsible for the production and breakdown of vitamins, growth hormones and phytotoxic substances (Gray and Williams 1971).

The plant ecologist is not usually concerned with the responses of microbiologically sterile plants, but it should be apparent from the above that the degree and type of nonsterility may radically affect experimental results and conclusions.

8.2.2 *Association of microflora*
with plant surfaces

Living plants provide a substrate for fungal and bacterial growth. Above-ground portions of plants, such as leaves, are only sparsely colonized except when senescent or diseased (Preece & Dickinson 1971) but roots typically show a characteristic microbial flora, often fungal (e.g. Parkinson et al. 1963). While such associations may influence the growth of plants, greater interest has attached to symbiotic relationships between plants and micro-organisms. Symbiotic nitrogen-fixation has already been mentioned (7.3.2) and various mycorrhizal associations may have a profound effect upon the physiology of the host plant (Meyer 1974). Mycorrhizas are usually divided into two basic types, ectotrophic where the fungus is largely external and endotrophic where the fungus is contained

TABLE 8.3 Growth and mineral content of mycorrhizal and non-mycorrhizal plants grown on poor soils.

	dry wt.	Nitrogen tot.	Nitrogen conc.	Phosphorus tot.	Phosphorus conc.	Potassium tot.	Potassium conc.	
	303	2·69	8·50	0·24	0·74	1·38	4·25	uninoculated
Pinus	405	5·00	12·40	0·79	1·96	3·02	7·44	inoculated
strobus	34	86	46	234	165	119	75	% increase
(ecto-)	93	1·17	7·58	0·24	1·56	0·98	6·20	uninoculated
	223	2·05	8·87	0·30	1·35	1·52	6·72	inoculated
	140	75	17	25	−13	55	8	% increase
Calluna	168	0·74	4·68	0·09	0·42	0·64	3·82	nonmycorrhizal
vulgaris	221	1·03	4·38	0·15	0·67	0·90	4·06	mycorrhizal
(endo-)								
	32	40	7	67	60	40	6	% increase

All weights in mg, tissue concentrations in mg g^{-1} (dry weight). (Data of Hatch 1937, Finn 1942, Bannister & Norton, 1974.)

with the cortical cells of the host. Ectotrophic mycorrhizas are typical of many trees and endotrophic mycorrhizas are often subdivided into orchid mycorrhizas, ericaceous mycorrhizas, and the vesicular-arbuscular types which appear to be very common. Recent work suggests that the different mycorrhizas have similar physiological effects. The fungus appears to rely on a supply of simple carbohydrate from the host (e.g. Lewis & Harley 1965, Smith *et al.* 1969, Stribley & Read 1974a) with the possible exception of orchid mycorrhizas where the converse is true (Smith 1966). On the other hand, the ectotrophic, ericaceous and vesicular-arbuscular mycorrhizas are able to accumulate phosphorus compounds (Harley 1969, Pearson & Read, 1973, Mosse *et al.* 1973) which may become available to the rest of the plant. The uptake of nitrogenous material may also be enhanced (Carrodus 1966, Bannister & Norton 1974, Stribley & Read 1974b) and there may also be effects on the uptake of other elements such as potassium (Harley & Wilson 1959, Bannister & Norton 1974). The net result of the enhanced uptake is usually an increase in growth (Table 8.3, Figure 8.4), particularly on poor soils although the beneficial effects of mycorrhizas may be lost or reversed on soils rich in nutrients (Mosse 1973, Bannister & Norton 1974).

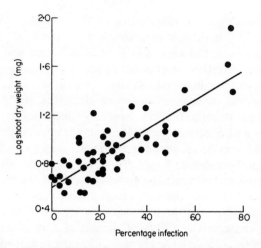

FIGURE 8.4. The relationship between the growth of *Lolium perenne* on a nutrient-poor soil and the degree of endomycorrhizal infection by *Rhizophagus tenuis* (Crush 1973).

8.2.3 *Disease*

Micro-organisms can also be agents of disease. Plants may stimulate the development of pathogens by their exudates but may also possess compounds, e.g. phenolics, which inhibit development or possess the capacity for producing inhibitory compounds (phytoalexins) as a response to infection (Kuć 1972).

Once established the pathogen disrupts the physiology of the host. Some, necrotrophic, parasites rapidly kill whole or part of the plaɪ and live saprophytically off the dead material; other, biotrophiↄ parasites have a more balanced relationship in which they conve.. their host's products into fungal materials. This situation is akin to that in mycorrhizas, except that the host derives no benefit. Necrotrophy is more opportunistic and is only likely to be of importance when some catastrophe upsets the balance in the community: biotrophy is more likely to be observed in a natural community of plants as it does not eliminate the host and can therefore persist. Thus leaves infected by biotrophic parasites show an increased translocation of sugars (Holligan *et al.* 1974) in infected leaves, plant hexoses are converted into fungal sugars such as trehalose (Long & Cooke 1974) and there is an increased incorporation of carbon-14 representing an increase in fungal lipids (Lösel & Lewis 1974).

Other effects upon the metabolism of plants include the vascular wilt diseases where there is controversy as to the relative effects of toxins and vascular blockage. Toxins such as fusaric acid may cause increased transpiration, decreased uptake of water and eventual death by desiccation in tomatoes (Gäumann 1958) whereas blockage by *Pseudomonas solanacearum* causes decreases in both photosynthesis and transpiration in bananas (Beckman *et al.* 1962). Respiratory and photosynthetic rates and enzyme, particularly oxidase, activity may be increased in the initial stages of infection but fall as tissue becomes damaged. Mildews may also affect photosynthetic rates, at least partially, by reducing the photosynthetic area. Increased metabolic activity may produce more food for the parasite but the final effects will be a decrease in vigour of the host. It is difficult to evaluate the effects of disease in natural communities where infection is likely to be chronic rather than epidemic. The loss of competitive vigour in an infected species must affect the balance of species in a community, whereas death due to disease is most likely to occur in species which are living at the limits of their tolerance range; such plants are likely to be killed when weakened rather than slowly die naturally. For example, plants grown in deep shade are likely to have thin cell walls and low sugar contents: both these modifications may predispose plants to infection (Grime 1966).

8.3   INTERACTIONS WITH ANIMALS

Animals affect the ecology of plants by acting as pollinators (Proctor & Yeo 1973) and as agents or dispersal. Trampling and manuring by large animals (see 8.4.3) may alter the habitat and alter the composition of plant communities, but the most frequent interaction between animals and plants, which has both physiological and ecological implications, is the potential use of plants as a food source. This interaction is considered in this section.

8.3.1   *The consumption of plants by animals*

Palatability can confer hardly any selective advantage on a plant. The reduction of leaf area and drain on resources that occur when a plant is eaten or recovering from being eaten must reduce competitive vigour. Accordingly many plants have evolved defence mechanisms such as hairiness, hardness and the presence of spines which deter many predators; unfortunately animals also evolve and have the advantage of being able to show changed patterns of behaviour (Dethier 1970) and thus even the best protected plants have their predators. Extreme forms of biochemical defence include plants that produce analogues of insect hormones which can potentially disrupt the development of feeding larvae (Williams 1970) but a more common defence is found in plants which are poisonous or foul tasting. Mild poisoning and repulsive tastes are likely to be more effective protection than fatal poisoning as then the animal can come to recognize the particular plant. Plants contain a large number of noxious substances; many of them incidental by-products of metabolism that have not evolved specifically for protection against being eaten but which may also have additional uses, such as in chemical antagonism between plants (8.1.2). These chemicals have a wide range of effects including addiction, photosensitization and carcinogenesis which are of dubious selective advantage as they either encourage consumption or produce delayed effects which are not readily associated with their source (Table 8.4).

TABLE 8.4. Poisonous plants and their effects. (Selected from Forsyth 1968).

Plant	Toxic substance	Effect
Cherry laurel	prussic acid	respiratory
(*Prunus laurocerasus*)	(from glycoside)	poison
Dog's mercury	mercurialine	chronic irritation of
(*Merurialis perennis*)	(volatile oil)	digestive tract
Ragwort	jacobine (and	toxic
(*Senecio jacobea*)	other alkaloids)	hepatitis
Opium poppy	morphine (alkaloid)	sleepiness
(*Papaver somniferum*)		addiction
Castor oil	ricin (phytotoxin)	convulsion and
(*Ricinus communis*)		death
St. John's Wort	hypericin (pigment)	photosensitization
(*Hypericum perforatum*)		
Bracken	unknown	carcinogenesis
(*Pteridium aquilinusm*)		

However, unpalatable plants, such as ragwort, *Senecio jacobea*, are avoided by grazing animals, whether they are relatively non-selective feeders like cattle or selective like sheep. It is interesting to

note that the caterpillar of the cinnabar moth (*Callimorpha jacobeae*
L.) consumes ragwort as its principal food and incorporates the
alkaloids in its tissues thus rendering itself unpalatable to its predators (Alpin *et al.* 1968); Caterpillars of the related burnet moths
(*Zygaena* spp.) have an analagous defence mechanism in that they
accumulate cyanide from cyanogenic trefoils (Jones *et al.* 1962). Other
invertebrates avoid certain plant extracts; for example the snail,
Cepaea nemoralis L., avoids filter paper treated with extracts of
plants such as *Hedera helix* and *Caltha palustris* and disfavours
other extracts (e.g. those from *Plantago lanceolata, Ranunculus acris*
and *Lathyrus pratensis*). When presented with a large range of plant
leaves, snails ate only one-fifth of the species presented and favoured
dicotyledons and those with soft leaves (Grime *et al.* 1968). Cyanogenic forms of clover (*Trifolium repens*) and other legumes (e.g.
Lotus corniculatus) may be avoided by herbivorous invertebrates
(e.g. Jones 1962). The cyanogenic forms are more frequent at low
altitudes where grazing populations are likely to be higher, their
scarcity at higher altitudes is possibly due to low winter temperatures
causing the release of cyanide in cyanogenic plants with the consequent inhibition of respiration (Daday 1954). This simple explanation has been questioned (Bishop & Korn 1969).

The chemical constitution of a plant may not only inhibit feeding,
it may stimulate it. Butterfly larvae are often associated with certain
food plants, this may be partly conditioning, but cabbage white
caterpillars (*Pieris brassicae* L.) reared on an artificial diet increase
their food intake when mustard glycosides are added (David &
Gardiner 1966). Higher animals also exhibit food preferences.
Captive ptarmigan (*Lagopus mutus*) are partly conditioned by their
previous diet but appear to select food of high nutritive value (Moss
1968) while heather shoots in the crops of grouse (*Lagopus lagopus
scoticus*) show a higher nitrogen and phosphorus content than shoots
collected randomly from the field (Moss 1969). When nitrogenous
fertilizers are added to small plots of heather, these areas are
preferentially grazed not only by grouse but also by hares and
rabbits (Miller 1968).

Grazing represents a loss of plant production and grazed plants
expend energy in regrowth and are obliged to take up new supplies
of minerals from the soil. Thus continued grazing will result in a
diminution of vigour, especially on unproductive sites. Consequently,
most plants of pastures possess adaptations that minimize the harm-

ful effects of grazing and the most common of these is the hemi-cryptophyte habit that allows plants to become closely cropped without damage to their growing points. Grazing may not be wholly deleterious and may help to keep plants in a physiologically young state; for example, the mineral content of artificially clipped heather shoots is higher than in unclipped plants (Grant & Hunter 1966). Chronic infestations of invertebrate pests may have even more dramatic effects upon growth. Normal infestations of the lime aphid are estimated to cause at least 19 per cent net energy drain on mature trees (Llewellyn 1975), leaf size in spring is reduced more than would be expected by the removal of nutrients alone, leaves are shed in autumn while still rich in nitrogen and root growth is inhibited (Dixon 1971b). Similar effects are found in sycamore (Dixon 1971a) and in both species the rate of dry matter production in infested leaves is greater than in uninfested leaves. This suggests that the materials are retained or even directed towards infested leaves, a situation comparable to that in leaves infested by obligate fungal parasites (8.2.3).

8.4 INTERACTIONS WITH MAN

Man, by his very nature, has altered and exploited his environment, and consequently rendered it less suitable for its former vegetation than for some other. Thus the elimination of forest cover restricted the forest flora to suitably shaded habitats such as copses, scrub, ravines and hedgerows whereas the open land was made suitable for pasture, arable crops and the weeds that were adapted to colonizing disturbed ground. The physiological relationships with the changed environments are the same as those described in previous chapters because man has not changed the fundamental physiological nature of plants, although he may have selected plants with particular physiological characteristics.

However, man has produced totally alien environments through both urban industry and rural agriculture. He has produced air with large amounts of suspended matter of diverse chemical nature, and added exotic chemicals to the land and water. Many plants have been eliminated from these spoiled habitats and only tolerant ones have survived. The physiological plant ecologist should be able to examine this tolerance and thus provide a rationale for the selection and breeding of plants able to tolerate these new environments. He can also examine the deleterious effects and give recommendations for acceptable levels of pollution.

8.4.1 *Agriculture*

In his agricultural activities man has not altered the basic physiology of plants but has selected and developed plants and plant characteristics which were of use to him. Thus the wild cabbage, *Brassica oleracea*, has yielded cultivars in kale, cabbage, brussel sprouts, kohl rabi, cauliflower and broccoli by modifications of leaves, buds, stems and inflorescences. These plants obviously are physiologically different but the potential for such development was inherent in their wild ancestors. Their relationships with the environment can be examined in the same way as any wild plant.

One of the differences between ecology and agriculture is that while mere survival can represent ecological success the agriculturalist is concerned with yield. Selective breeding can redirect assimilates to the appropriate part of the plant, as in the cabbage cultivars, but the agriculturalist is also concerned with altering both the plant and its habitat in order to increase yield. Plants can be made more efficient by increasing their ability to utilize solar energy; thus the agriculturalist has been concerned both with extending the growing season of crops (e.g. Robson & Jewiss 1968) and in increasing their leaf area index. For example, grasses from Mediterranean latitudes are adapted to winter growth and provide the potential for breeding strains of grass that would give both an early and late 'bite' in northern temperature regions. One of the disadvantages of moving plants from their latitude of origin is that their phenological development may be out of phase with the seasonal progression in their new environment. Consequently they may be susceptible to frost and drought if these occur at different times of year, and the temperature and photoperiodic regimes may alter the course of their development. The investigation of such phenomena is, in effect, 'applied' physiological ecology. It is interesting to note that man in the tropics has unwittingly cultivated a number of grasses with C_4 metabolism (e.g. maize, sugarcane), thus benefitting from their efficient photosynthesis under high light intensity.

Yield may also be increased by modifying the soil environment, particularly, by the addition of fertilizers. Some of the new cultivars, e.g. the Mexican wheats, produce their high yields only when supplied with large amounts of fertilizer. Good cultivation also adds to yield by promoting root growth, aeration, humification, water-holding and the retention and recirculation of nutrients. On the negative side the release of nutrients from cultivated land by leaching (in addition to inputs from both treated and untreated sewage) leads to the eutrophication of waters and consequent pollution, a topic of considerable contemporary concern (e.g. Ehrlich & Ehrlich 1972).

A still further contribution to increased yield is made by the elimination of competitors and pests. Modern agriculture relies upon chemical herbicides (Fryer & Makepeace 1972) and pesticides—these represent new factors in the environment and thus plant responses and tolerances to them are of some interest.

Herbicides are usually classified as total, when they kill all plants, and selective, when they kill only certain plants. In total herbicides there is an advantage in a combination of high toxicity with rapid breakdown. This allows replanting of cleared areas soon after application; in areas that are to be kept clear there is an advantage in persistence but low mobility so that other areas are not affected. Paraquat fulfils the first need and herbicides such as simazine the second. Despite their drastic nature, total herbicides may not always be completely effective; for example, paraquat is less effective on nitrogen-deficient cereal leaves (Lutman *et al.* 1975).

Selective herbicides kill only a limited spectrum of plants. Theoretically, selectivity could be based on biochemical differences between plants, but usually morphological and anatomical differences provide the necessary basis. Consequently broad-leaved species are susceptible to the translocated, 'hormonal' weed-killers such as 2, 4-D largely because their horizontally placed leaves retain and absorb the herbicide whereas the more vertically oriented grass leaves with their silicified surfaces both allow the chemical to drain away and are resistant to its absorbtion. A species may show variability in its resistance to herbicide on a seasonal basis (Figure 8.5); during the growing season plants showing reduced growth and stunted form are usually more resistant than vigorously growing ones.

Susceptibility may also be associated with growth form; ragwort, *Senecio jacobea*, is less resistant in its first year of growth when it has a rosette form than when it is erect and flowering in its second year. Different forms of the same species may also have different

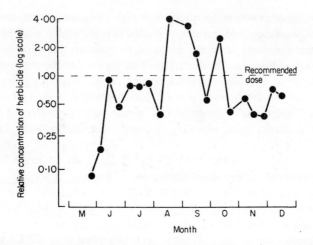

FIGURE 8.5. Seasonal changes in the resistance of shoots of *Calluna vulgaris* to 2,4-D ester. (Data of Scott 1971).

Resistance is estimated as the proportion of the recommended dose (equivalent to 156 ppm) required to induce an estimated 20 per cent damage.

These are laboratory determinations under standard conditions; resistance in the field in governed by prevailing temperatures and weather and plants are more susceptible in the warmer months.

resistances as been shown for both the field thistle, *Cirsium arvense*, (Hodgson 1964) and bindweed (Whitworth & Muzik 1967).

Pesticides that are not designed for use against higher plants may nevertheless affect them. Compounds of copper and mercury are common in fungicides and their accumulation in soils could produce toxic concentrations, while their continued application could lead to the selection of tolerant forms. Copper fungicides have long been used in vineyards, and there is some evidence for a higher copper tolerance in vines than in other cultivated plants (Repp 1963).

Insecticides, particularly those based on organo-chlorine compounds, have been the cause of much ecological concern because of their widespread dispersal and their concentration along food chains, resulting in a death and infertility of wildlife. The various case histories are well known and need not be repeated here (see Ehrlich & Ehrlich 1972, Mellanby 1967). Their effects on plants are less frequently considered, particularly as the removal of insect pests usually results in enhanced yield, but the demonstration that extremely low concentrations of DDT can reduce the photosynthesis of phytoplankton (Figure 8.6) is a matter of considerable concern. As more and more agricultural chemicals are produced, there will be a continuing need to monitor their physiological and ecological effects upon plants and other organisms.

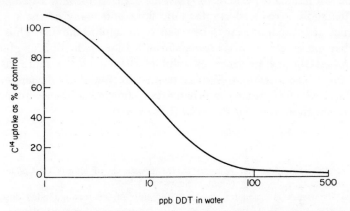

ppb DDT in water

FIGURE 8.6. Effect of DDT upon the photosynthesis of marine phytoplankton in water from Vineyard Sound, Woods Hole, Mass. (Redrawn from Wurster 1968).

8.4.2 Industry

Increased industrial production is concomitant with the increased production and accumulation of waste products. Some of these are exotic, but many are merely naturally occurring chemicals which occur in abnormal quantities. Mining and smelting produce areas contaminated with metals and the reactions of plants to metalliferous soils has already been considered (7.3.3). The reclamation of these and other contaminated sites usually involves the alleviation of any toxicity, the adjustment of pH and the correction of nutrient imbalances. The response of planted species to various treatments can be examined experimentally (e.g. Fitter & Bradshaw 1974a,b).

Industry has also contaminated the aerial environment; smoke may reduce the amount of light available for photosynthesis and particles may cover leaves and clog stomata. On the other hand, increased carbon dioxide and temperature levels might even offset some of the loss of photosynthetic ability that occurs in urban environments. Arguably, the greatest effect man has had on the composition of the atmosphere is the production of large amounts of sulphur dioxide as a by-product of burning fossil fuels which are almost always contaminated with sulphur. When the gas combines with atmospheric water it produces sulphurous acid and sulphite, and further oxidization may produce sulphuric acid and sulphates. The last named is perhaps the least harmful. In Britain, urban areas often show a reduced cryptogamic flora and many of the species which are absent appear to be sensitive to sulphur dioxide. Thus the urban moss, *Bryum argenteum*, shows a lesser reduction of photosynthesis in the presence of sulphite than does the more rural, *Hypnum cupressiforme*. Similar results are found for lichens with *Usnea,* which is missing in urban areas, showing a greater reduction than the more tolerant *Parmelia* (Figure 8.7).

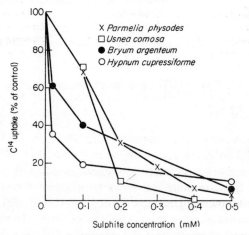

FIGURE 8.7. Influence of sulphite on the photosynthesis of lichens and mosses (data of Hill 1974 and Inglis & Hill 1974).

Higher plants are also affected by sulphur dioxide, yield may be reduced (Bell & Clough 1973), probably because of the action of sulphur dioxide on photosynthetic enzymes (e.g. Ziegler 1972); although other processes may be affected, as in beans where stomatal closure is impaired, thus rendering the plant susceptible to desiccation (Biscoe *et al.* 1973). Large and continued doses of sulphur dioxide will ultimately lead to the death of susceptible plants. Resistance has been examined by fumigating a range of species with the gas, the relative resistance thus obtained can provide a basis for the selection of species suitable for growth in urban areas (Table 8.5).

TABLE 8.5. Relative resistance of woody plants to sulphur dioxide (abstracted from Ranft & Dässler 1971).

Very sensitive	Sensitive	
Hypericum calycinum	Larix leptolepis*	
Larix decidua*	Pinus montana	
Picea abies	Potentilla fruticosa	
Pinus sylvestris	Rubus idaeus	
Salix purpurea	Salix fragilis*	
	Tilia cordata	

Intermediate hardiness	Hardy	Very resistant
Alnus glutinosa	Erica carnea	Buxus sempervirens
Cornus alba	Ilex aquifolium	Chamaecyparis pisifera
Crataegus monogyna	Hedera helix	Ligustrum vulgare
Fagus sylvatica	Prunus padus	Lonicera periclymenum
Fraxinus excelsior	Sambucus niger*	Platanus acerifolia
Sarothamnus scoparius	Thuja plicata	Quercus petraea
Syringa vulgaris		

Species marked thus* were shown to be able to produce vigorous regrowth after a period of acute exposure.

There are many more examples of industrial pollution, and it is not intended to catalogue them here. Rather, the examples given should show that physiological-ecological investigations of pollution may follow the same course as the investigation of any other 'natural' environmental factor. The correlation of vegetation and environment leads to relevant physiological experiments and the determination of resistance and the limits of tolerance of species.

8.4.3 Recreation

The increased personal mobility which has occurred in the post-war period has lead to excessive pressure on certain areas of countryside. Much of this is literal pressure — exerted by numerous feet. The damage that ensues is physical rather than physiological in that plants are deformed, crushed and broken. This results in the replacement of sensitive forms by more resistant types. In chalk grassland, herbaceous species such as Poterium sanguisorba, Leontodon hispidus, Trifolium repens and the moss, Rhytidiadelphus squarrosus all tend to disappear; whereas the most frequent species in trampled areas include *Carex flacca*, *Plantago lanceolata* and *Bellis perennis*, while the grass, *Dactylis glomerata* is common in both trampled and unaffected areas (Chappell *et al.* 1971). In upland areas, species such as the deer sedge, *Trichophorum caespitosum*, are more resistant than dwarf shrubs such as *Calluna vulgaris* (Bayfield 1971). In general, it seems that tough nonwoody plants with growing points near to the soil surface are favoured hence the prominence of grassy species and herbaceous plants with rosette habits. Woody plants, which at first might be thought to be resistant, are not because their stems are easily broken and their growing points exposed and unprotected. Ecotypic differentiation may occur and *Plantago lanceolata* produces more prostrate ecotypes in trampled areas (Mrs J. Watson, unpublished data).

The resistance of species may be quantified by artificial trampling. Such studies have shown that timothy grass, *Phleum pratense*, is actually stimulated by light trampling although heavier pressures result in decreased performance. Light pressure probably encourages tillering (Bayfield 1971).

Recreational activities may also alter the habitat. Eutrophication may occur through inputs of litter, urine and faeces in places visited by large numbers of humans with their attendant pets. A popular area of chalk grassland was found to lack some of the characteristic species (e.g. *Thymus drucei*, *Asperula cynanchica*) although some were more resistant (e.g. *Medicago lupulina*). Grasses, such as *Cynosurus cristatus* and *Lolium perenne* became dominant and the increase in *Lolium* appeared to be related to increased phosphorus and nitrogen in the soil (Streeter 1971).

8.5   RELATIONSHIPS OF LABORATORY AND FIELD STUDIES

The preceding text should have made it evident that while experiment and physiological investigation may help to interpret a plant's distribution in the field, they do not necessarily provide a complete explanation and in some cases may provide none at all. The main reason for this is that experiments are often concerned with a limited number of variables whereas the plant responds to a complex of interacting variables. Interpretation is made easy only when one or two variables have an overriding influence.

Physiological experiments are usually made with plants that are well grown. Light, space, water, nutrients and temperatures are usually at favourable levels, except when one of these influences is under investigation. In the field, any or all of the controlling influences may be at a suboptimal level. Thus, in some cases, amelioration of any factor may be observed to have a beneficial effect. For example, developing tree seedlings in shade may show increased growth, not only when shade is removed, but also by the addition of water and nutrients and by the removal of grazing pressures and root competition. This sort of situation is possibly more common than the 'limiting factor' situation, where the factor in the shortest supply limits the development of the plant and the amelioration of other factors would have little or no effect. Many experiments implicitly accept a limiting factor hypothesis and in consequence, when the alteration of a factor is shown to result in increased growth, it is often assumed to be the controlling factor, although the influence of other factors may not have been investigated.

There is also a temptation to consider that the most obvious environmental factor must be the one which controls distribution. Thus light intensity must govern the response of plants to sun and shade, calcium status (or at least soil chemistry) must determine the distribution of calcicoles and calcifuges and microclimatic factors must account for plants with differing altitudinal ranges. Such factors may well have an important influence, but it is worth remembering that, for example, plants from shade and those suffering from nutrient deficiency may have altered water relations and plants from high altitudes may be adapted to the absence of competition.

Thus, differences between experimental results obtained under controlled conditions and the responses of plants in the field may be partly a consequence of experimental philosophy and interpretation. However, the factor that causes the greatest difference between laboratory and field responses is probably the influence of other plants. This may be conveniently illustrated by a consideration of the responses of various heath plants to water regimes (Figure 8.8). In the field, *Erica cinerea*, *E. tetralix* and *Calluna* are characteristic of dry, wet and intermediate regimes respectively, whereas *Empetrum nigrum* may be found in both wet and dry sites. In the laboratory, *E. cinerea*, *E. tetralix*, and *Calluna* all show their best growth on soils of intermediate moisture status (Bannister 1964b) and the same is probably true of *Empetrum*. However, *Calluna* is a vigorous

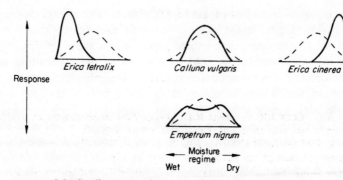

FIGURE 8.8. A diagrammatic comparison of ecological (———) and physiological (– – –) amplitude of four heathland dwarf shrubs with respect to soil moisture (Bannister 1976, based on Ellenberg 1958 and Rorison 1969).

competitor (Gimingham 1972) and ousts the other species to regimes where their physiological adaptations allow them to complete successfully with *Calluna*. Thus *Erica cinerea*, which is more drought tolerant than *Calluna*, can compete successfully on dry soils; *E. tetralix*, which is more tolerant of waterlogging, can compete successfully on wet sites; while *Empetrum nigrum* possibly has distinct ecotypes (Gimingham 1972) adapted to wet and dry regimes respectively. Similar patterns of response are shown for other species and other environmental variables (Ellenberg 1958, Rorison 1969).

In conclusion, therefore, this book has attempted to show that a consideration of physiological responses to specific environmental variables can provide at least partial explanations of ecological phenomena. Ultimately, however, the snippets of information thus obtained have to be put together to form a picture of the whole complex of plant-environment relationships which constitutes plant ecology. I hope that this book has gone a little way towards that end as well.

References

ALBEN A.O., COLE J.R. & LEWIS R.D. (1932). New developments in treating pecan rosette with chemicals. *Phytopathology* 22, 979–981

ALDEN J. & HERMANN R.K. (1971) Aspects of the cold-hardiness mechanism in plants. *Bot. Rev.* 37, 37–142

ALLEN R. & SHEPPARD P.M. (1971) Copper tolerance in some Californian populations of the monkey flower, *Mimulus guttatus. Proc. R. Soc. B* 172, 177–196

ALLISON R.V., BRYAN O.C. & HUNTER J.H. (1927) The stimulation of plant growth on the raw peat soils of the Florida Everglades through use of copper sulphate and other chemicals. *Fla. Agric. Stn. Bull.* 190

ALPIN R.T., BENN M.H. & ROTSCHILD M. (1968) Poisonous alkaloids in the body tissues of the cinnabar moth (*Callimorpha jocobeae* L.) *Nature, Lond.* 219, 747–748

ANDERSON M.C. (1964) Studies in the woodland light climate. I. The photographic computation of light conditions. *J. Ecol.* 52, 27–41

ANDERSON M.C. (1966a) Stand structure and light penetration. II. A theoretical analysis. *J. appl. Ecol.* 3, 41–54

ANDERSON, M.C. (1966b) Some problems of simple characterization of the light climate in plant communities. In: *Light as an ecological factor* (eds.R. Bainbridge, G.C. Evans & O. Rackham) pp. 77–90 Blackwell Scientific Publications, Oxford

ANTONOVICS J. (1966) The genetics and evolution of differences between closely adjacent plant populations with special reference to heavy metal tolerance. Ph.D. Thesis, University of Wales

ANTONOVICS J., BRADSHAW A.D. & TURNER R.G. (1971) Heavy metal tolerance in plants *Adv. ecol. Res.* 7, 1–85

ARENS J. (1939) Bestimmung des Turgordruckes an einer Einzelzelle mit dem Manometer. *Planta,* 30, 113–117

ATKINSON D. (1973) Observations on the phosphorus nutrition of two sand dune communities at Ross Links. *J. Ecol.* 61, 117–133

ATKINSON D. & Davison A.W. (1901) The effect of phosphorus deficiency on the growth of *Epilobium montanum* L. *New Phytol.* 70, 789–797

AUBERT B. (1971) Effets de la radiation globale sur la synthése d'acides organiques et la régulation stomatique des plantes. *Oecol. Pl.* 6, 24–34

BAKHUIS J.A. & KLETER H.J. (1965) Some effects of associated growth on grass and clover under field conditions. *Neth. J. agric. Sci.* 13, 280–310

BANNISTER P. (1963) The water relations of certain heath plants with reference to their ecological amplitude. Ph. D. thesis, University of Aberdeen

BANNISTER P. (1964a) Stomatal responses of heath plants to water deficits. *J. Ecol.* 52, 151–8

BANNISTER P. (1964 b,c,d) The water relations of certain heath plants with reference to their ecological amplitude.
   I. Introduction, germination and establishment. *J. Ecol.*, **52**, 423–432
   II. Field studies. *J. Ecol.* **52**, 468–481
   III. Experimental studies: General conclusions. *J. Ecol.* **52**, 499–510

BANNISTER P. (1965) Biological Flora of the British Isles. *Erica cinerea* L. *J. Ecol.* **53**, 527–542

BANNISTER P. (1970) The annual course of drought and heat resistance in heath plants from an oceanic environment. *Flora, Jena,* **159**, 105–23

BANNISTER P. (1971) The water relations of heath plants from open and shaded habitats. *J. Ecol.* **59**, 61–64

BANNISTER P. (1973) A note on some observations on frost damage in the field, with particular reference to various ferns. *Trans bot. Soc. Edinb.* **42**, 111–113

BANNISTER P. (1976) Physiological ecology and plant nutrition. In:*Methods in plant ecology* (ed. S. B. Chapman), pp. 229–295. Blackwell Scientific Publications, Oxford

BANNISTER P. & NORTON W.M. (1974) The response of mycorrhizal and non-mycorrhizal rooted cuttings of heather (*Calluna vulgaris* (L.) Hull) to variations in nutrient and water regime. *New Phytol.* **73**, 81–90

BARBER D.A. & LOUGHMAN, B.C. (1967) The effect of micro-organisms on the absorbtion of mineral nutrients by intact plants. II. Uptake and utilisation of phosphate by barley plants grown under sterile and non-sterile conditions· *J. exp. Bot.* **18**, 170–176

BARBER D.A. (1969) The influence of the microflora on the accumulation of ions by plants. In: *Ecological aspects of the mineral nutrition of plants.* (ed. I.H. Rorison) pp. 191–200. Blackwell Scientific Publications, Oxford

BARBER D.A., EBERT M. & EVANS N.T.S. (1962) The movement of ^{15}O through barley and rice plants. *J. exp. Bot.* **13**, 397–403

BARTHOLEMEW B. (1970) Bare zone between California shrub and grassland communities: the role of animals. *Science, N.Y.* **170**, 1210–1212

BAUMGARTNER A. (1953) Das Eindringen des Lichtes in dem Boden. *Forstwiss. ZentBl.* **72**, 172–184

BAUMGARTNER A. (1955) Licht und Naturverjüngung am Nordrand eines Waldbestandes. *Forstwiss. ZentBl.* **74**, 59–64

BAUMGARTNER A. (1956) Untersuchungen über den Wärme- und Wasserhaushalt eines junges Waldes. *Ber. dt. Wetterd., Bad Kissingen.* **5**, Nr. 28

BAYFIELD N.G. (1971) Some effects of walking and skiing on vegetation at Cairngorm. In: *The scientific management of animal and plant communities for conservation* (eds. E. Duffey & A.S. Watt). pp. 469–485, Blackwell Scientific Publications, Oxford

BECKMANN C.H., BRUN, W.A. & BUDDENHAGEN I.W. (1962) Water relations of banana plants infected with *Pseudomonas solanacearum*. *Phytopathology* **53**, 1144–1148

BEHRE K. (1929) Physiologische und zytologische Untersuchungen über *Drosera*. *Planta* **7**, 208–306

BIEBL R. (1964) Temperaturresistenz tropischer Pflanzen auf Costa Rica. *Protoplasma* **59**, 133–156

BELL J.N. & CLOUGH W.S. (1973). Repression of yield in ryegrass exposed to
SO₂. *Nature, Lond.* **241**, 47–49

BERGH J.P. VAN DEN & ELBERSE W. Th. (1962) Competition between *Lolium
perenne* L. and *Anthoxanthum odoratum* L. at two levels of phosphate and
potash. *J. Ecol.* **50**, 87–95

BERGH J.P. VAN DEN (1969) Distribution of pasture plants in relation to chemical
properties of the soil. In: *Ecological aspects of mineral nutrition of plants*
(ed. I.H. Rorison) pp. 11–23. Blackwell Scientific Publications, Oxford

BILLINGS W.D. & MORRIS R.J. (1951) Reflection of visible and infra-red radiation
from leaves of different ecological groups. *Am. J. Bot.* **38**, 327–331

BINET P. (1968) Dormances et aptitude a germer en milieu sale chez les Halophytes. *Bull Soc. fr. Physiol. Vég.* **14**, 115–124

BIRKEBAK R. & BIRKEBAK R. (1964) Solar radiation characteristics of tree leaves.
Ecology **45**, 646–649

BISCOE P.V., UNSWORTH M.H. & PINKSEY H.R. (1973) The effects of SO₂ on
stomatal behaviour in *Vicia faba. New Phytol.* **72**, 1299–1306

BISHOP J.A. & KORN M.E. (1969) Natural selection and cyanogenisis in white
clover, *Trifolium repens. Heredity, Lond.* **24**, 423–430

BJÖRKMAN O. (1971) Comparative photosynthetic CO₂ exchange in higher
plants. In: *Photosynthesis and photorespiration* (eds. M.D. Hatch, C.B.
Osmond & R.O. Slayter) pp. 18–32. Wiley-Interscience, New York

BJÖRKMAN O. & BERRY J. (1973) High efficiency photosynthesis. *Sci. Am.* **229**(4),
80–93

BLACK C.C. (1971) Ecological implications of dividing plants into groups with
distinct photosynthetic production capacities. *Adv. ecol. Res.* **7**, 87–114

BLACK M. (1969) Light controlled germination of seeds. *Symp. Soc. Exp. Biol.*
23, 193–217

BLACK M. & WAREING P.F. (1955) Photoperiodic control of germination in
Betula pubescens Ehrh. *Physiol. Pl.* **8**, 300–316

BLACK M. & WAREING P.F. (1960). Photoperiodism in the light inhibited seed
of *Nemophila insignis. J. exp. Bot.* **11**, 28–39

BLACK R.F. (1960) Effects of NaCl on the ion uptake and growth of *Atriplex
vesicaria* Howard. *Aust. J. Biol. Sci.* **13**, 249–266

BLACKMAN G.E. & RUTTER A.J. (1948) Physiological and ecological studies in
the analysis of plant environment. III. The interaction between light intensity
and the mineral nutrient supply in leaf development and in the net assimilation
rate of the bluebell (*Scilla non-scripta*). *Ann. Bot.* N.S. **12**, 1–26

BLACKMAN G.E. & BLACK J.N. (1959) Physiological and ecological studies in the
analysis of plant environment. XI. A further assessment of the influence of
shading on the growth of different species in the vegetative phase. *Ann. Bot.*
N.S. **23**, 51–63

BOATMAN D.J. (1972) The growth of *Schoenus nigricans* on blanket bog peats.
II. Growth on Irish and Scottish peats. *J. Ecol.* **60**, 469–477

BODE H.R. (1958) Beiträge zur Kenntnis allelopathischer Erscheinungen bei
einigen Juglandaceen. *Planta* **51**, 440–480

BOND G. (1967) Fixation of nitrogen by higher plants other than legumes.
Ann. Rev. Plant Physiol. **18**, 107–126

BOND G. & SCOTT G.D. (1955) An examination of symbiotic systems for fixation of nitrogen. *Ann. Bot.* N.S. **19**, 67–77

BÖRNER H. (1960) The liberation of organic substances from higher plants and their role in the soil sickness problem. *Bot. Rev.* **26**, 393–424

BOWMAN G.E. (1968) The measurement of carbon dioxide concentration in the atmosphere. In: *The measurement of environmental factors in terrestrial ecology* (ed. R.M. Wadsworth) pp. 131–140. Blackwell Scientific Publications, Oxford

BRADSHAW A.D. (1969) The ecologist's viewpoint. In: *Ecological aspects in the mineral nutrition of plants* (ed. I.H. Rorison) pp. 415–427. Blackwell Scientific Publications, Oxford

BRAUN-BLANQUET J. (1932) *Plant Sociology: the study of plant communities* (English Translation) McGraw-Hill, New York

BRIX H. (1962) The effect of water stress on the rates of photosynthesis and respiration in tomato plants and loblolly pine seedlings. *Physiol. Pl.* **15**, 10–20

BROWN I.C. (1943) A rapid method of determining exchangeable hydrogen and total exchangeable bases in soils. *Soil Sci.* **56**, 353–357

BROUWER, R. (1953) Water absorbtion by the roots of *Vicia faba* at various transpiration strengths. I. Analysis of the uptake and the factors determining it. *Proc. K. ned. Akad. Wet.* C**56**, 106–115

BROUWER R. (1954) Water absorbtion by the roots of *Vicia faba* at various transpiration strengths. III. Changes in water conductivity artificially obtained. *Proc. K. ned. Akad. Wet.* C**57**, 68–80

BRUNT D. (1932) Notes on radiation in the atmosphere. *Q. Jl. R. met. Soc.* **58**, 389–418

BUCHSBOM U. (1968) Salzresistenz oberirdischer Sprossteile von Holzgewächsen. I. Chlorideinwirkung aut Sprossoberflächen. *Flora, Jena.* **157**, 527–561

BUCHSBOM U. (1969) Salzresistenz oberirdischer Sprossteile von Holzgewächsen. II. Chlorideinwirkung auf die Achsengewebe—Jahreslauf der Resistenz. *Flora, Jena* **158**, 129–158

BULL T.A. (1971) The C_4 pathway related to growth rates in sugarcane. In: *Photosynthesis and photorespiration* (eds. M.D. Hatch, C.B. Osmond & R.O. Slatyer) pp. 68–75. Wiley-Interscience, New York

BÜSGEN M. (1883) Die Bedeutung des Insektenfanges für *Drosera rotundifolia.* *Bot. Ztg.* **41**, 569–577; 585–594

CAPALETTI PAGANELLI E.M. (1968) L'alofitismo e le sue variazioni stagionali in alcune specie litorali. *Webbia* **23**, 101–133

CARRODUS B.B. (1966) Absorbtion of nitrogen by mycorrhizal roots of beech. I. Factors affecting the assimilation of nitrogen. *New Phytol.* **65**, 358–371

CAVERS P.B. & HARPER J.L. (1964) Biological Flora of the British Isles: *Rumex crispus* L. *J. Ecol.* **52**, 754–766

CHAPMAN S.B. (1976) *Methods in Plant Ecology.* Blackwell Scientific Publications, Oxford

CHAPPELL H.G., AINSWORTH J.F., CAMERON R.A.D. & REDFERN M. (1971) The effect of trampling on a chalk grassland ecosystem. *J. appl .Ecol.* **8**, 869–881

CHOUARD, P. (1946) Sur le photoperiodisme chez les plantes vivaces. *Bull soc. bot Fr.* **93**, 373–77

CLARKSON D.T. (1965) Calcium uptake by calcicole and calcifuge species in the genus *Agrostis*. *J. Ecol.* **53**, 427–435

CLARKSON D.T. (1966) Aluminium tolerance within the genus *Agrostis*. *J. Ecol.* **54**, 167–178

CLARKSON D.T. (1967) Phosphorus supply and growth rates in species of *Agrostis* L. *J. Ecol.* **55**, 707–731

COLLIS-GEORGE N. & SANDS J.E. (1959) The control of seed germination by moisture as a soil physical property. *Aust. J. agric. Res.* **10**, 628–636

COLLMANN W. (1958) Diagramme zum Strahlungsklima Europas. *Ber. dt. Wetterd., Bad Kissingen* **6**, Nr. 42

COUPLAND R.T. & JOHNSON R.E. (1965) Rooting characteristics of native grassland species in Saskatchewan. *J. Ecol.* **53**, 475–507

COWAN I.R. & MILTHORPE F.L. (1968) Plant factors influencing the water status of plant tissues. In: *Water deficits and plant growth* (ed. T.T. Kozlowski) Vol. I. pp. 137–193. Academic Press, New York

CRAFTS A.S., CURRIER H.B. & STOCKING C.R. (1949) *Water in the physiology of plants*. Chronica Botanica Co., Waltham, Mass.

CRAIG G.C. (1972) The population genetics of *Agrostis canina* with respect to lead tolerance. M.Sc. Thesis, University of Stirling

CRAWFORD R.M.M. (1966) The control of anaerobic respiration as a determining factor in the distribution of the genus *Senecio*. *J. Ecol.* **54**, 403–413

CRAWFORD R.M.M. (1972) Physiologische Ökologie: Ein Vergleich der Anpassung von Pflanzen und Tieren an sauerstoffarme Umgebung. *Flora, Jena* **161**, 209–233

CRAWFORD R.M.M. & TYLER P.D. (1969) Organic acid metabolism in relation to flooding tolerance in roots. *J. Ecol.* **57**, 237–246

CRUSH J.R. (1973) The effects of *Rhizophagus tenuis* mycorrhizas on rye grass, cocksfoot and sweet vernal. *New Phytol.* **72**, 965–973

CUMMING B.G. (1963) Dependence of germination on photoperiod, light quality and temperature in *Chenopodium* spp. *Can. J. Bot.* **41**, 1211–1233

CURTIS J.T. & McINTOSH R.P. (1951) An upland forest continuum in the prairie-forest border region of Wisconsin. *Ecology* **32**, 476–496

DADAY, H. (1954) Gene frequencies in wild populations of *Trifolium repens* L. II. Distribution by altitude. *Heredity, Lond.* **8**, 377–384

DAUBENMIRE R.F. (1972) Phenology and other characteristics of tropical semideciduous forest in north western Costa Rica. *J. Ecol.* **60**, 147–170

DAUBENMIRE R.F. (1974) *Plants and Environment* (3rd edn). Wiley, New York

DAVID W.A.L. & GARDINER B.O.C. (1966) Mustard oil glucosides as feeding stimulants for *Pieris brassicae* larvae in a semi-synthetic diet. *Entomologia exp. appl.* **9**, 247–255

DAVIES M.S. & SNAYDON R.W. (1973,1974) Physiological differences among populations of *Anthoxanthum odoratum* L. collected from the Park Grass Experiment, Rothamsted
  I. Response to calcium. *J. appl. Ecol.* **10**, 33–45
  II. Response to aluminium. *J. appl. Ecol.* **10**, 47–55

III. Response to phosphorus. *J. appl. Ecol.* **11**, 699–707

DAVISON A.W. (1970) The ecology of *Hordeum murinum* L. I. Analysis of the distribution in Britain. *J. Ecol.* **68**, 453–466

DETHIER V.G. (1970) Chemical interactions between plants and insects. In: *Chemical Ecology* (ed. E. Sondheimer & J.B. Simeone) pp. 83–102. Academic Press, London & New York

DIXON A.F.G. (1971a,b) The role of aphids in wood formation.
  I. The effect of the sycamore aphid, *Drepanosiphum platanoides* (Schr) (Aphididae) on the growth of sycamore, *Acer pseudoplatanus* (L.). *J. appl. Ecol.* **8**, 165–179
  II. The effect of the lime aphid, *Eucallipterus tilia* L. (Aphididae) on the growth of lime, *Tilia* × *vulgaris* Hayne. *J. appl. Ecol.* **8**, 393–399

DONEEN, L.D. & McGILLIVRAY J.R. (1943). Germination (emergence) of vegetable seeds as affected by different soil moisture conditions. *Pl. Physiol., Lancaster* **18**

DOWNTON W.S. (1971) Check list of C_4 species. In: *Photosynthesis and photorespiration* (eds. M.D. Hatch, C.B. Osmond & R.O. Slatyer) pp. 554–558. Wiley-Interscience, New York

EATON F.M. (1942) Toxicity and accumulation of chloride and sulphate in plants. *J. agric. Res.* **64**, 357–399

EATON F.M. & ERGLE D.R. (1948) Carbohydrate accumulation in the cotton plant at low moisture levels. *Pl. Physiol., Lancaster* **23**, 169–187

ECKERSON S. (1913) A physiological and chemical study of after-ripening *Bot. Gaz.* **55**, 286–299

EIJK M. VAN (1939) Analyse der Wirkung des NaCl auf die Entwicklung, Sukkulenz und Transpiration bei *Salicornia herbacea* sowie Untersuchungen über den Einfluss der Salzaufnahme auf die Wurzelatmung bei *Aster tripolium*. *Extr. Rec. Trav. bot. Neerl.* **36**, 559–567

EKERN P.C. (1965) Evapotranspiration of pineapple in Hawaii. *Pl. Physiol., Lancaster* **40**, 736–739

ELLENBERG H. (1958) Mineralstoffe für die pflanzliche Besiedlung des Bodens. A. Bodenreaktion (einschliesslich Kalkfrage). *Handb. Pl. Physiol.* IV., 638–709

EHRLICH P.R. & EHRLICH A.H. (1972) *Population, resources, environment.* (2nd edn.) Freeman, San Fransisco

ERNST, W. (1965a) Ökologische-Soziologische Untersuchungen der Schwermetallpflanzengesellschaften Mitteleuropas unter Einschluss der Alpen. *Abh. Landesmus. Naturk. Münster* **27**, 1–54

ERNST W. (1965b) Über den Einfluss des Zinks auf die Keimung von Schwermetallpflanzen und auf die Entwicklung der Schwermetallpflanzengesellschaft. *Ber. dt. bot. Ges.* **78**, 205–212

ERNST W. (1968a) Der Einfluss der Phosphatersorgung sowie die Wirkung von ionogenem und chelatisiertem Zink auf die Zink- und Phosphataufnahme einiger Schwermetallpflanzen. *Physiol. Pl.* **21**, 323–333

ERNST W. (1968b) Ökologische Untersuchungen an Pflanzengesellschaften unterschiedlich stark gestörter Böden in Grossbrittanien. *Flora, Jena* **158**, 95–109

EPSTEIN E. (1969) Mineral metabolism of halophytes. In: *Ecological aspects of the mineral nutrition of plants* (ed. I.H. Rorison) pp. 345–355. Blackwell Scientific Publications, Oxford

ETHERINGTON J.R. (1975) *Environment and plant ecology.* Wiley-Interscience, London

EVANS G.C. (1956) An area survey method of investigating the distribution of light intensity in woodlands, with particular reference to sunflecks. *J. Ecol.* **44**, 391–428

EVANS G.C. (1965) Model and measurement in the study of woodland light microclimate. In: *Light as an ecological factor* (ed. R. Bainbridge, G.C. Evans & O. Rackham). Blackwell Scientific Publications, Oxford

EVANS G.C. (1972) *The quantitative analysis of plant growth.* Blackwell Scientific Publications, Oxford

EVANARI M. (1965) Light and seed dormancy. *Handb. Pl Physiol.* **XV/2**, 804–847

FINN R.G. (1942) Mycorrhizal inoculation of soil of low fertility. *Black Rock For. Pap.* **1**, 115–117

FIRBAS F. (1932) Untersuchungen über den Wasserhaushalt der Hochmoorpflanzen. *Jb. wiss. Bot.* **74**, 459–696

FITTER A.H. & BRADSHAW A.D. (1974a) Responses of *Lolium* and *Agrostis* to reclamation treatments. *J. appl. Ecol.* **11**, 597–608

FITTER A.H. & BRADSHAW A.D. (1974b) Root penetration of *Lolium perenne* on colliery shale in response to reclamation treatments. *J. appl. Ecol.* 608–615

FLINT H.L. (1972) Cold hardiness of twigs of *Quercus rubra* L. as a function of geographic origin. *Ecology* **53**, 1163–1170

FORSYTH A.A. (1968) *British poisonous plants.* M.A.F.F. Bull. No. 161, HMSO, London

FRYER J.D. & MAKEPEACE R.J. (1972) *Weed Control Handbook* (7th edn). Vol. I. Principles. Vol. II. Recommendations. Blackwell Scientific Publications, Oxford

GABRIELSEN E.K. (1948) Effects of different chlorophyll concentration on photosynthesis in foliage leaves. *Physiol. Pl.* **1**, 5–37

GAFF D.F. (1966) The sulfhydryl-disulphide hypothesis in relation to desiccation injury of cabbage leaves. *Aust. J. Biol. Sci.* **19**, 291–299

GAFF D.F. & CARR D.J. (1964) An examination of the refractrometric method for determining the water potential of plant tissues. *Ann. Bot.* N.S. **28**, 352–368

GARDNER W.R. (1960) Dynamic aspects of water availability to plants. *Soil Sci.* **89**, 63–73

GARDNER W.R. & NIEMANN R.H. (1964) Lower limit of water availability to plants. *Pl. Physiol., Lancaster* **40**, 705–710

GATES D.M. (1962) *Energy exchange in the biosphere.* Harper & Row, New York

GATES D.M. & TANTRAPORN W. (1952) The reflectivity of deciduous trees and herbaceous plants in the infra-red to 25 microns. *Science, N.Y.* **115**, 613–616

GARTSIDE D.W. & McNEILLY T. (1974a,b) Genetic studies in heavy metal tolerant plants.

I. Genetics of zinc tolerance in *Anthoxanthum odoratum Heredity, Lond.* 32, 287–297

II. Zince tolerance in *Agrostis tenuis. Heredity, Lond.* 33, 303–308

GARTSIDE D.W. & MCNEILLY T. (1974c) The potential for evolution of heavy metal tolerance in plants. II. Copper tolerance in normal population of different plant species. *Heredity, Lond.* 32, 335–348

GÄUMANN E. (1958) The mechanism of fusaric acid injury. *Phytopathology* 48, 67–686

GEIGER R. (1964) *The climate near the ground* (Translation of the German 4th edn). Harvard University Press, Cambridge Mass.

GEIGER R. & AMANN H. (1932) Forstmeteorologische Messungen in einem Eichenbestand. *Forstwiss. ZentBl.* 54, 371–383

GIGON A. & RORISON I.H. (1972) The response of some ecologically distinct plant species to nitrate and ammonium nitrogen. *J. Ecol.* 60, 93–102

GIMINGHAM C.H. (1960) Biological Flora of the British Isles. *Calluna vulgaris* (L.) Hull. *J. Ecol.* 48, 455–483

GIMINGHAM C.H. (1949) The effects of grazing on the balance between *Erica cinerea* and *Calluna vulgaris* in upland heath and their morphological responses. *J. Ecol.* 37, 100–119

GIMINGHAM C.H. (1960) Biological Flora of the British Isles. *Calluna vulgaris* (L.) Hull. *J. Ecol.* 48, 455–483

GIMINGHAM C.H. (1972) *The ecology of heathlands.* Chapman & Hall, London

GIMINGHAM C.H. & CORMACK E. (1964). Plant distribution and growth in relation to aspect on hill slopes in N. Scotland. *Trans. bot. Soc. Edinb.* 39, 525–538

GITTINS R. (1969) The application of ordination techniques. In: *Ecological aspects of the mineral nutrition of plants* (ed. I.H. Rorison), pp. 37–66. Blackwell Scientific Publications, Oxford

GÖHRE K. & LÜTZKE R. (1956) Der Einfluss von Bestandsdichte und -struktur auf das Kleinklima im Walde. *Arch. Forstw.* 5, 487–572

GOLDSMITH F.B. (1973) The vegetation of exposed sea cliffs at South Stack, Anglesey. II. Experimental Studies. *J. Ecol.* 61, 819–829

GORHAM E. (1959) A comparison of lower and higher plants as accumulators of radioactive fallout. *Can. J. Bot.* 37, 327–329

GRADMANN H. (1928) Untersuchungen über die Wasserverhältnisse des Bodens als Grundlage des Pflanzenwachstums. *Jb. wiss. Bot.* 69, 1–100

GRANT S.A. & HUNTER R.F. (1962) Ecotypic differentiation of *Calluna vulgaris* in relation to altitude. *New Phytol.* 61, 44–56

GRANT S.A. & HUNTER R.F. (1966) The effects of frequency and season of clipping on the morphology, productivity and chemical composition of *Calluna vulgaris* (L.) Hull. *New Phytol.* 65, 125–133

GRAY T.R.G. & WILLIAMS S.T. (1971) *Soil micro-organisms,* Oliver & Boyd, Edinburgh

GREGORY R.P.G. & BRADSHAW A.D. (1965) Heavy metal tolerance in populations of *Agrostis tenuis* Sibth. and other grasses. *New Phytol.* 64, 131–143

GREEN F.H.W. (1962) Potential evaporation measurements. *British Rainfall 1958.* Pt. III, pp. 10–14. HMSO, London

GREEN F.H.W. (1964) The climate of Scotland. In: *The vegetation of Scotland* (ed. J.H. Burnett) pp. 15–35. Oliver & Boyd, Edinburgh

GRIME J.P. (1963) An ecological investigation at a junction between two plant communities in Coombsdale on the Derbyshire limestone. *J. Ecol.* **51**, 391–402

GRIME J.P. (1966) Shade avoidance and shade tolerance in flowering plants. In: *Light as an ecological factor* (ed. R. Bainbridge, G.C. Evans & O. Rackham) pp. 187–207. Blackwell Scientific Publications, Oxford

GRIME J.P. & JEFFREY D.W. (1965) Seedling establishment in vertical gradients of sunlight. *J. Ecol.* **53**, 621–642

GRIME J.P., MACPHERSON-STEWART S.F. & DEARMAN R.S. (1968) An investigation of leaf palatability using the snail, *Cepea nemoralis* L. *J. Ecol.* **56**, 405–420

GRIME J.P. & HODGSON J.G. (1969) An investigation of the ecological significance of lime-chlorosis by the means of large-scale comparative experiments. In: *Ecological aspects of the mineral nutrition of plants* (ed. I.H. Rorison) pp. 67–99. Blackwell Scientific Publications, Oxford

GRIME J.P. & LLOYD P.S. (1973) *An ecological atlas of grassland plants.* Arnold, London

GRIME J.P. & HUNT R. (1975) Relative growth-rate: its range and adaptive significance in a local flora. *J. Ecol.* **63**, 393–422

GRUBB P.J., GREEN H.E. & MERRIFIELD R.C.J. (1969) The ecology of chalk heath: its relevance to the calcicole-calcifuge and soil acidification problems. *J. Ecol.* **57**, 175–212

GRÜMMER G. & BEYER H. (1960) The influence exerted by species of *Camelina* on flax by means of toxic substances. In: *The biology of weeds* (ed. H.L. Harper) pp. 153–164. Blackwell Scientific Publications, Oxford

GRUNOW J. (1952) Beiträge zum Hangklima. *Ber. dt. Wetterd. U.S. Zone* **5(35)**, 293–298

GRUNOW J. (1955) Nebelniederschlag. *Ber. dt. Wetterd, U.S. Zone* **7(42)**, 30–34

GRUNDON, N.J. (1972) Mineral nutrition of some Queensland heath plants. *J. Ecol.* **60**, 171–181

GUNARY D. & SUTTON C.D. (1967) Soil factors affecting plant uptake of phosphate. *J. Soil Sci.* **18**, 167–173

HACKETT C. (1965) Ecological aspects of the nutrition of *Deschampsia flexuosa* (L.) Trin. II. The effects of Al, Ca, Fe, K, Mn, P and pH on the growth of seedlings and established plants. *J. Ecol.* **53**, 315–333

HAECKEL E. (1869) Über Entwicklungsgang und Aufgabe der Zoologie. *Jena Z. Naturw.* **5**, 353–370

HAINES F.M. (1928) A method of investigating and evaluating drought resistivity and the effect of drought conditions upon water economy. *Ann. Bot.* **42**, 667–705

HANDLEY W.R.C. (1963) Mycorrhizal associations and *Caluna* heathland afforestation. *Bull. For. Commn. London* **36**, 1–70

HARDER R. (1921) Kritische Versuche zu Blackmans Theorie der 'begrenzenden Faktoren' bei der Kohlensäureassimilation. *Jb. wiss. Bot.* **60**, 531–611

HARLEY J.L. (1969) *The biology of mycorrhiza* (2nd edn.) Leonard Hill, London

HARLEY J.L. & WILSON J.M. (1959) The absorption of potassium by beech mycorrhiza. *New Phytol.* **58**, 281–298

HARPER J.L. (1967) The teaching of experimental plant ecology. In: *The teaching of ecology* (ed. J. M. Lambert) pp. 135–145 Blackwell Scientific Publications, Oxford

HARPER J.L. & SAGAR G.R. (1953) Some aspects of the ecology of buttercups in permanent grassland. *Proc. Br. Weed Control Conf.* **1**, 256–265

HARPER J.L. & BENTON R.A. (1966) The behaviour of seeds in soil II The germination of seeds on the surface of a water supplying substrate *J. Ecol.* **54**, 151–166

HARPER J.L. & OGDEN J. (1970) The reproductive strategy of higher plants I The concept of strategy with special reference to *Senecio vulgaris* L. *J. Ecol.* **58**, 681–698

HATCH, A. B. (1937) The physical basis of mycotrophy in *Pinus strobus Black Rock For. Bull.* **6**, (168pp)

HATCH M.D. SLACK C.R. & JOHNSON H.S. (1967) Further studies on a new pathway of photosynthetic carbon dioxide fixation in sugar cane and its occurrence in other plant species. *Biochem. J.* **102**, 417–422

HEATH O.V.S. (1969) *The physiological aspects of photosynthesis.* Heinemann, London

HEATH O.V.S. & MEIDNER H. (1967) Compensation points and carbon dioxide enrichment for lettuce grown under glass in winter *J. exp. Bot.* **18**, 746–751

HELLMUTH E.O. (1968, 1969) Eco-physiological studies on plants in arid and semi-arid regions in Western Australia
  I. Autecology of *Rhagodia baccata* (Labil) Moq. *J. Ecol.* **56**, 319–344
  II. Field physiology of *Acacia craspedocarpa* F. Muell *J. Ecol.* **57**, 613–634

HELLMUTH, E.O. & GRIEVE, B.J. (1969) Measurement of water potential of leaves with particular reference to the Schardakow method. *Flora, Jena,* **159**, 147–167

HESLOP-HARRISON Y. & KNOX R.B. (1971) A cytochemical study of leaf-gland enzymes of insectivorous plants of the genus *Pinguicula. Planta* **96**, 183–211

HEWITT E.J. & SMITH T.A. (1975) *Plant mineral nutrition.* English Universities Press, London

HILL, D.J. (1974) Some effects of sulphite on photosynthesis in lichens. *New Phytol.* **73**, 1193–1215

HODGSON J.M. (1964) Variations in ecotypes of canada thistle. *Weeds* **12**, 167–70

HOFMANN G. (1955) Die Thermodynamik der Taubildung. *Ber. dt. Wetterd., Bad Kissingen,* **3**, Nr. 18

HOLLIGAN P.M. CHEN C. MCGEE E.E.M. & LEWIS D.H. (1974) Carbohydrate metabolism in healthy and rusted leaves of coltsfoot. *New Phytol.* **73**, 881–888

HOLMGREN, P. JARVIS P.G. & JARVIS M.S. (1965) Resistances to carbon dioxide and water vapour transfer in leaves of different plant species. *Physiol. Pl.* **18**, 557–573

HONERT T.H. van den (1948) Water transport as a catenary process *Discuss. Faraday Soc.,* **3**, 146–53

HORAK O. & KINZEL H. (1971) Typen des Mineralstoffwechsels bei den höheren Pflanzen. *Öst. bot. Z.* **119**, 475–495

HORI T. (1953) *Studies on fog*. Tanne Trading Co., Sapporo, Japan

HORTON R.E. (1919) Rainfall interception. *Month. Weath. Rev.* 47, 603–623

HUBER B. (1935) *Die Wärmehaushalt der Pflanzen.* Datterer, Munich

HUBER B. (1956) *Die Saftströme der Pflanzen.* Springer, Berlin

HUBER, B. & MERKENSCHLAGER G. (1951) Über die Wirkungen kleinste Saugkräfte auf die Samenkeimung. *Planta*, 40, 112–120

HUGHES A.P. (1965) Plant growth and the aerial environment. VI The apparent efficiency of conversion of light energy of different spectral composition by *Impatiens parviflora. New Phytol.* 64, 48–54

HUGHES A.P. & FREEMAN P.R. (1967) Growth analysis using frequent small harvests. *J. appl. Ecol.* 4, 553–560

HUNT R. & PARSONS I.T. (1974) A computer program for deriving growth functions in plant growth analysis. *J. appl. Ecol.* 11, 297–307

HUNTER A.C.J. (1971) Salt tolerance in *Festuca rubra* B.A. (Hons) Thesis, University of Stirling

HUTCHINSON T.C. (1967, 1970a, b) Lime chlorosis as a factor in seedling establishment on calcareous soil I A comparative study of species from acidic and calcareous soils in their susceptibility to lime chlorosis. *New Phytol.* 66, 697–705
II. The development of leaf water deficit in plants showing lime chlorosis *New Phytol.* 69, 143–157
III. The ability of green and chlorotic plants fully to reverse large leaf water deficits. *New Phytol.* 69, 261–268

HUTCHINSON T.C. (1968) A physiological study of *Teucrium scorodonia* ecotypes which differ in their susceptibility to lime-induced chlorosis and to iron deficiency chlorosis. *Pl. Soil.* 28, 81–105

HYGEN G. (1953) On the transpiration decline of excised plant samples. *Norske Vid. Akad. Skr. l. math. nat. Kl.* 1, 1–84

IDLE D.B. (1970) The calculation of transpiration rate and diffusion resistance of a single leaf from micrometeorological information subject to errors of measurement. *Ann. Bot.* N.S. 34, 159–176

INGLIS F. & HILL D.S. (1974) The effect of sulphite and fluoride in CO_2 uptake by mosses in the light. *New Phytol.* 73, 1207–1213

ISJKAWA S. & YOKOHAMA Y. (1962) Effect of 'intermittent irradiation' on the germination of *Epilobium* and *Hypericum Bot. Mag., Tokyo* 75, 127–132

IVIMEY-COOK R.B. & PROCTOR M.C.F. (1966) The plant communities of the Burren, Co. Clare. *Proc. R. Irish Acad.* 64B, 211–301

JARVIS M.S. (1963) A comparison between the water relations of species with contrasting types of geographical distribution in the British Isles. In: *The water relations of plants* (ed. A. J. Rutter & F. H. Whitehead) pp. 289–312 Blackwell Scientific Publications, Oxford

JARVIS P.G. (1964) The adaptability to light intensity of seedlings of *Quercus petraea* (Matt.) Liebl. *J. Ecol.* 52, 545–571

JARVIS P.G. (1965) Interference by *Deschampsia flexuosa* L. Trin. *Oikos,* 15, 56–78

JARVIS P.G. & JARVIS M.S. (1963a). Effect of several osmotic substrates on the growth of *Lupinus albus* seedlings. *Physiol. Pl.* 16, 485–500

JARVIS P.G. & JARVIS M.S. (1963b, c) The water relations of tree seedlings.
I. Growth and water use in relation to soil water potential. *Physiol. Pl.*
16,
IV. Some aspects of the tissue water relations and drought resistance *Physiol.
Pl.* **16,** 501–516

JARVIS P.G. & JARVIS M.S. (1964) Growth rates of woody plants *Physiol. Pl.* **17,**
654–666

JEFFERIES R.L. & WILLIS A.J. (1964) Studies of the calcicole-calcifuge habit.
II. The influence of calcium on the growth and establishment in soil and sand
culture. *J. Ecol.* **52,** 691–707

JEFFERIES R.L. LAYCOCK D. STEWART G.R. & SIMS A.P. (1969) The properties
and mechanisms involved in the uptake and utilisation of calcium and
potassium by plants in relation to an understanding of plant distribution.
In: *Ecological aspects of the mineral nutrition of plants* (ed. I. H. Rorison)
pp. 281–307 Blackwell Scientific Publications, Oxford

JEFFREY D.W. (1964) The formation of polyphosphate in *Banksia ornata*, an
Australian heath plant. *Aust. J. Biol. Sci.* **17,** 845–854

JEFFREY D.W. (1968) Phosphate nutrition of Australian heath plants. II. The
formation of polyphosphate by five heath species. *Aust. J. Bot.* **16,** 603–
613

JEFFREY D.W. (1970) A note on the use of ignition losses as a means for the
approximate estimation of soil bulk density. *J. Ecol.* **58,** 297–299

JEFFREY D.W. (1971) The experimental alteration of a *Kobresia*-rich sward in
Upper Teesdale. In: *The scientific managment of animal and plant communities
for conservation* (ed. E. Duffey & A. S. Watt) pp. 79–89 Blackwell Scientific
Publications, Oxford

JEFFREY D.W. & PIGOTT C.D. (1973) The response of grasslands on sugar
limestone to applications of phosphorus and nitrogen. *J. Ecol.* **16,** 85–92

JENNINGS D.H. (1968) Halophytes, succulence and sodium in plants—a unified
theory. *New Phytol.* **67,** 899–911

JENSEN C.R., STOLZY L.H. & LETEY J. (1967) Labelled oxygen transport through
growing corn roots. *Soil Sci.* **,103** 23–29

JENSEN R.D.S. TAYLOR S.A. & WIEBE H.H. (1961) Negative transport and
resistance to water flow through plants. *Pl. Physiol.* Lancaster, **36,** 633–
638

JOHNSTON W.R. (1974) Mineral uptake of plants from serpentine and lead mine
soils B.A. (Hons.) Thesis, University of Stirling

JONES D.A. (1962) Selective eating of the acyanogenic form of the plant, *Lotus
corniculatus* L. by various animals. *Nature. Lond.,* **193,** 1109

JONES D.A. PARSONS J. & ROTHSCHILD M. (1962) Release of hydrocyanic acid
from crushed tissues of all stages of the life cycles of the Zygaeninae (Lepi-
doptera) *Nature, Lond.* **193,** 52–53

JONES H.E. (1971a, b) Comparative studies in plant growth and distribution in
relation to waterlogging
II. An experimental study of the relationship between transpiration and the
uptake of iron in *Erica cinerea* L. and *E. tetralix* L. *J. Ecol.* **59,** 167–178
III. The response of *Erica cinera* to waterlogging in peats of different iron
content *J. Ecol.* **59,** 583–591

JONES H.E. & ETHERINGTON J.R. (1970) Comparative studies of plant growth and distribution in relation to waterlogging. I. The survival of *Erica cinerea* L. and *Erica tetralix* L. and its apparent relationship to iron and manganese uptake in waterlogged soil *J. Ecol.* **58**, 487–496

JOST L. (1906) Über die Reaktionsgeschwindigkeit im Organismus. *Biol. Zentbl.* **26**, 225–244

KAEMPFERT W. (1942) Sonnenstrahlung auf Ebene, Wand und Hang *Wiss. Abh. Reichsamt WettDienst.* **9**(3)

KAEMPFERT W. & MORGEN A. (1952) Die Besonnung *Z. Met.* **6**, 138–146

KAPPEN, L. (1966) Der Einfluss des Wassergehaltes auf die Widerstandfähigkeit von Pflanzen gegenüber hohen und tiefen Temperaturen, untersucht an Blättern einiger Farne und von *Ramonda myconi. Flora Jena* **156**, 427–445

KAUFMAN M.R. (1968) Water relations of pine seedlings in relation to root and shoot growth. *Pl. Physiol., Lancaster* **43**, 281–288

KAUSCH W. (1952) Physiologische Wirkungen kleinster Saugkräfte. *Planta* **41**, 59–63

KELLER R. & CLODIUS S. (1956) Schema des Wasserkreislaufes für Gebiet der Bundesrepublik Deutschlands und das Jahr 1951 *Der Grosse Herder* **9**, 906

KERSHAW K.A. (1973) *Quantative and Dynamic Ecology* (2nd edn) Arnold, London

KNIPLING E.B. (1967) Effects of ageing on water deficit-water potential relationships of dogwood leaves growing in two environments. *Physiol. Pl.* **20**, 65–72

KOHNKE H. (1968) *Soil Physics* McGraw-Hill, New York

KOLLER D. (1969) The physiology of dormancy and survival of plants in desert envionments. In: *Dormancy and Survival* (ed. H. W. Woolhouse) *Symp. Soc. exp. Biol.* **23**, 449–469

KOZLOWSKI, T.T. (1965) *Water metabolism in plants* Harper & Row, New York

KREEB K. (1960) Über die gravimetrische Methode zur Bestimmung der Saugspannung und das Problem des negativen Turgors. *Planta,* **55**, 274–282

KREEB K. (1963) Hydrature and plant production. In: *The water relations of plants.* (ed. A. J. Rutter & F. H. Whitehead) pp. 272–288, Blackwell Scientific Publications, Oxford

KREEB, K. (1974) *Ökophysiologie der Pflanzen.* Fischer, Jena

KRENN K. (1933) Die Bestrahlungsverhältnisse stehender und liegender Stämme. *Wien. allg. Forst-Jagdztg.* **51**, 50–54

KUĆ J. (1972) Phytoalexins *Ann. Rev. Phytopathol* **10**, 207–232

LANGE O.L. (1959) Untersuchungen über den Wärmehaushalt und Hitzeresistenz mauretanischer Wüsten und Savannenpflanzen. *Flora Jena* **147**, 598–651

LANGE O.L. (1961) Die Hitzeresistenz einheimscher immer- und wintergrüner Pflanzen im Jahreslauf. *Planta* **56**, 666–683

LANGE O.L. (1962) Versuche der Hitzeresistenz-Adaptation bei höheren Pflanzen. *Die Naturwiss.* **49**, 20–21

LANGE O.L. (1969) Die funktionellen Anpassungen der Flechten an die ökologischen Bedingungen arider Gebiete. *Ber. dt. bot. Ges.* **82**, 3–22

LANGE O.L. & Schwemmle B. (1960) Untersuchungen zur Hitzeresistenz vegativer und blühenden Pflanzen von *Kalanchoë blossenfeldiana. Planta* **55**, 208–255

LANGE O.L. & LANGE R. (1962) Die Hitzeresistenz einiger mediterraner Pflanzen in Abhängigkeit von der Höhenlage ihrer Standorte. *Flora, Jena* **152**, 707–710

LANGE O.L. & LANGE R. (1963) Untersuchungen über Blatttemperaturen, Transpiration und Hitzeresistenz an Pflanzen mediterraner Standorte (Costa Brava, Spanien) *Flora, Jena* **153**, 387–425

LANGE O.L. & KANZOW H. (1965) Wachstumshemmung an höheren Pflanzen durch abgetöte Blätter und Zwiebeln von *Allium ursinum. Flora, Jena* **156** 94–101

LANGE O.L. KOCH W. & SCHULZE E.D. (1969) CO_2 gas exchanges and water relationships of plants in the Negev Desert at the end of the dry period. *Ber. dt. bot. Ges.* **82**, 39–61

LARCHER W. (1969) The effect of environmental and physiological variables on the carbon dioxide gas exchanges of trees. *Phytosynthetica* **3**, 167–198

LARCHER W. (1971) Die Kälteresistenz von Obstbaümen und Ziergehölzen subtropischer Herkunft. *Oecol. Pl.* **6**, 1–14

LARCHER W. (1973a) *Ökologie der Pflanzen*. Ulmer, Stuttgart

LARCHER W. (1973b) Limiting temperatures for life functions in plants. In: *Temperature and life* (2nd edn.) (eds. H. Precht, J. Christophersen, H. Hensel & W. Larcher) Springer, Berlin

LARCHER W. & MAIR B. (1968) Das Kälteresistenzverhalten von *Quercus pubescens, Ostyra carpinifolia* und *Fraxinus ornus* auf drei thermisch unterschiedlichen Standorten *Oecol. Pl.* **3**, 255–270

LARSEN S. & SUTTON C.D. (1963) The influence of soil volume on the absorbtion of soil phosphorus by plants and on the determination of labile phosphorus. *Pl. Soil.* **81**, 77–84

LARSON P.R. (1964) Some indirect effects of environment on wood formation. In: *The functions of wood in trees* (ed. M. Zimmerman) pp. 345–365 Academic Press, New York

LAWRENCE D.B. SCHOENIKE A. QUISPEL A. & BOND G. (1967) The role of *Dryas drummondii* in vegetation development following ice recession at Glacier Bay, Alaska, with special reference to its nitrogen fixation by root nodules. *J. Ecol.* **55**, 793–813

LAZENBY A. (1955) Germination and establishment of *Juncus effusus* L. II. The interaction effects of moisture and competition *J. Ecol.* **43**, 595–605

LEE J.A. & WOOLHOUSE H.W. (1969) A comparative study of bicarbonate inhibition of root growth in calcicole and calcifuge grasses. *New Phytol.* **68** 1–14

LEVITT J. (1956) *Hardiness in plants*. Academic Press, New York

LEVITT J. (1958) Frost, drought and heat resistance. *Protoplasmatologia* VIII **6**, Springer, Vienna.

LEVITT J. (1962) A sulfhydryl-disufide hypothesis of frost injury and resistance in plants. *J. theor. Biol.* **3**, 355–391

LEVITT J. (1972) *Responses of plants to environmental stresses*. Academic Press, New York

LEVITT J. SULLIVAN C.Y. & KRULL E. (1960) Some problems in drought resistance *Bull. Res. Council, Israel*, **8D**, 173–180

LEWIS D.H. & HARLEY J.L. (1965) Carbohydrate physiology of mycorrhizal roots of beech III. Movement of sugars between host and fungus. *New Phytol.* **64**, 256–269

LEWIS, M.C. (1972) The physiological significance of variation in leaf structure *Sci. Prog. Lond.* **60**, 25–51

LEYST E. (1890) Über die Bodentemperaturen in Pawlowsk. *Reprium Met. St. Petersb.* **13**, Nr 7, 1–31

LIETH H. (1950) Grenzen und Anwendungsmöglichkeiten der colorimetrischen CO_2 Bestimmung. *Planta* **51**, 705–721

LLEWLLYN M. (1975) The effects of the lime aphid, *Eucallipterus tilliae* L., (Aphididae) on the growth of lime (*Tilia × vulgaris* Hayne) II. The primary production of saplings and mature trees, the energy drain imposed by the aphid population and revised standard deviations of aphid population energy budgets. *J. Appl. Ecol.* **12**, 15–23

LLOYD P.S. GRIME J.P. & RORISON I.H. (1971) The grassland vegetation of the Sheffield region I. General features *J. Ecol.* **59**, 863–886

LONG, D.E. & COOKE R.C. (1974) Carbohydrate composition and metabolism of *Senecio squalidus* leaves infected with *Albugo tragopogonis* (Pers.) S. F. Gray *New Phytol* **73**, 889–899

LÖSEL D.M. & LEWIS D.H. (1974) Lipid metabolism in leaves of *Tussilago farfara* during infection by *Puccinia poarum*. *New Phytol.* **73**, 1157–1169

LOUW H.A. & WEBLEY D.M. (1959) A study of soil bacteria dissolving certain phosphatic mineral fertilizers and related compounds. *J. appl. Bact.* **22**, 227–233

LUTMAN P.J.W. SAGAR G.R. MARSHALL C. & HEADFORD D.W.R. (1975) The influence of nitrogen status on the susceptibility of segments of cereal leaves to paraquat. *Weed Res.* **15**, 89–92

MCGILCHRIST C.A. & TRENBATH B.R. (1971) A revised analysis of plant competition experiments. *Biometrics* **27**, 659–671

MCKELL C.M., PERRIER E.R. & STEBBINS G.L. (1960) Response of two subspecies of orchard grass (*Dactylis glomerata*, subspecies *lusitanica* and *judaica*) to increasing soil moisture stress. *Ecology* **41**, 772–778

MCPHERSON J.K. & MULLER C.H. (1969) Allelopathic effects of *Adenostoma fasiculatum*, 'Chamise', in the Californian chaparral. *Ecol. Monogr.* **39**, 177–198

MARTIN M.H. & PIGOTT C.H. (1965) A simple method for measuring carbon dioxide in soils. *J. Ecol.* **53**, 153–156

MARTIN P. & RADEMACHER B. (1960) Studies on the mutual influences of weeds and crops. In: *The Biology of weeds* (ed. J.L. Harper) pp. 143–153, Blackwell Scientific Publications, Oxford

MATTHEWS S. (1971) A study of seed lots of peas (*Pisum sativum* L.) differing in predisposition to pre-emergence mortality in soil. *Ann. appl. Biol.* **68**, 177–183

MAXIMOV N.A. (1929) *The plant in relation to water* (English trans.). Allen & Unwin, London

MEIDNER H. (1962) The minimum intra-cellular- CO_2 concentration. *J. exp. Bot.* **13**, 284–293

MEIDNER H. & MANSFIELD T.A. (1968) *Physiology of stomata*. McGraw-Hill, London

MEIDNER H. & EDWARDS M. (1975) Pressure potentials in guard cells. *Nature, London.* **253**, 114–115

MEIDNER H. & WILLMER C. (1975) Mechanics and metabolism of guard cells. *Curr. Adv. Pl. Sci.* **7**, 1–15

MELLANBY K. (1967) *Pesticides and pollution.* Collins, London

MEYER B.S. (1945) A critical evaluation of the terminology of diffusion phenomena. *Pl. Physiol., Lancaster* **20**, 142–162

MEYER F.H. (1974) Physiology of mycorrhiza. *Ann. Rev. Pl. Physiol.* **25**, 567–586

MICHAEL G. (1967) Über die Beanspruchung des Wasserhaushaltes einiger immergrüner Gehölze im Mittelgebirge in Zusammenhang mit dem Frost-trocknisproblem. *Arch.Forstw.* **16**, 1015–1032

MILBURN J.A. (1970) Cavitation and osmotic potential of *Sordaria* ascospores. *New Phytol.* **69**, 133–144

MILLINER L.H. (1962) Daylength and *Ulex europaeus* L. II. Ecotypic variation with latitude. *New Phytol.* **61**, 119–128

MILLER G.R. (1968) Evidence for selective feeding on fertilized plots by red grouse hares and rabbits. *J. Wildl. Mgmt.* **32**, 849–853

MITCHELL F. (1973) Roadside plant responses to lead and sodium. B.A.(Hons) Thesis, University of Stirling

MOLISCH H. (1937) *Der Einfluss einer Pflanze auf die Andere.* Fischer, Jena

MONSI M. & SAEKI T. (1953) Über den Lichtfaktor in den Pflanzengesellschaften und seine Bedeutung für die Stoffproduktion. *Jap. J. Bot.* **14**, 22–52

MONTEITH J.L. (1963) Dew: Facts and fallacies. In: *The water relations of plants.* (eds. A.J. Rutter & F.H. Whitehead) pp. 37–56. Blackwell Scientific Publications, Oxford

MONTEITH J.L. (1973) *Principles of environmental physics.* Arnold, London

MONTGOMERY E.G. (1912) Competition in cereals. *Bull. Nebr. agric. exp. Stn.* **26**, *Art. V.* 22 pp.

MOONEY H.A. & BILLINGS W.D. (1961) Comparative physiological ecology of arctic and alpine populations of *Oxyria digyna*. *Ecol. Monogr.***31**, 1–29

MOONEY H.A. & SHROPSHIRE F. (1968) Population variability in temperature related photosynthetic capacities. *Oecol. Pl.* **2**, 1–13

MOSS R. (1968) Food selection and nutrition in Ptarmigan (*Lagopus mutus*). *Symp. Zool. Soc., London* **21**, 207–216

MOSS R. (1969) Nutrition in red grouse and ptarmigan. In: *Grouse Research in Scotland.* 13th Progress Report. pp. 18–24, Nature Conservancy, Edinburgh

MOSS R.A. & LOOMIS W.E. (1952) Absorbtion spectra of leaves. I. The visible spectrum. *Pl. Physiol., Lancaster* **27**, 370–391

MOSSE B. (1973) Plant growth response to vesicular-arbuscular mycorrhiza. IV. In soil given added phosphate. *New Phytol.* **72**, 127–136

MOSSE B., HAYMAN D.S. & ARNOLD D.J. (1973) Plant growth in response to vesicular-arbuscular mycorrhiza. V. Phosphate uptake by three plant species from P-deficient soils labelled with ^{32}P. *New Phytol.* **72**, 809–815

MULLER W.H. (1965) Volatile materials produced by *Salvia leucophylla*. Effects on seedling growth and soil bacteria. *Bot. Gaz.* **126**, 195–200

MULLER W.H. & HAUGE R. (1967) Volatile growth inhibitors produced by *Salvia leucophylla*: effect on seedling anatomy. *Bull. Torrey bot. Club.* **91**, 327–330

MÜLLER-STOLL W.R. & LERCH G. (1963) Model tests on the ecological effect of vapour movement and condensation in soil due to temperature gradients. In: *Water relations of plants* (eds. A.J. Rutter & F.H. Whitehead) pp. 65–82. Blackwell Scientific Publications, Oxford

MULQUEEN J., WALSHE M.J. & FLEMING G.A. (1961) Copper deficiency on Irish blanket peats. *Sci. Proc. R. Dubl. Soc.* B1, 25–35

MYERSCOUGH P.J. & WHITEHEAD F.H. (1966) Comparative biology of *Tussilago farfara, Chamaenerion angustifloium, Epilobium montanum* L. and E. *adenocaulon* Huasskn. I. General biology and germination. *New Phytol.* 65, 192–210

NASSERY H. & HARLEY J.L. (1969) Phosphate absorbtion by plants from habitats of different phosphate status. I. Absorbtion and incorporation of phosphate by excised roots. *New Phytol.* 68, 13–20

NEWMAN E.I. (1966) Relationship between root growth of flax (*Linum usitatissimum*) and soil water potential. *New Phytol.* 65, 273–283

NEWTON J.E. & BLACKMAN G.E. (1970) The penetration of solar radiation through canopies of different structure. *Ann. Bot.* 34, 329–348

NORMAN J.T., KEMP A.W. & TAYLER J.E. (1957) Winter temperatures in long and short grass. *Meteorol. Mag.* 86, 148–152

NYE P.H. (1969) The soil model and its application to plant nutrition. In: *Ecological aspects of the mineral nutrition of plants* (ed. I.H. Rorison) pp. 105–114. Blackwell Scientific Publications, Oxford

OKSBJERG E. (1966) On the shoot growth and susceptibility to frost of three conifers. *Hedeselsk. Tidsskr., Aarhus* 87, 359–371

OOURA H. (1953) The capture of fog by particles in the forest. *J. Met. Res. Tokyo* 4 (supplement), 239–259

ORDIN L. (1958) The effect of water stress on the cell wall metabolism of plant tissue. In: *Radioisotopes in scientific research.* Vol. 4 pp. 553–564. Pergamon, New York

ORDIN L. (1960) Effect of water stress on cell wall metabolism of *Avena* coleoptile tissue. *Pl. Physiol., Lancaster* 35, 443–450

OUDMAN J. (1936) Über Aufnahme und Transport N-hältiger Verbindungen durch die Blätter von *Drosera capensis. Ext. Rec. Trav. Bot. Neerl.* 33, 351–433

OVINGTON J.D. (1954) A comparison of rainfall in different woodlands. *Forestry* 27, 41–53

OVINGTON J.D. & LAWRENCE D.B. (1964) Strontium 90 in maize field, cattail marsh and oakwood ecosystems. *J. appl. Ecol.* 1, 175–181

OWEN P.C. (1952) The relation of germination of wheat to water potential. *J. exp. Bot.* 3, 188–203

PARK D. (1960) Antagonism—the background to soil fungi. In: *The ecology of soil fungi.* Liverpool University Press. pp. 148–159

PARKER J. (1969) Further studies of drought resistance in woody plants. *Bot. Rev.* 35, 317–371

PARKINSON D. TAYLOR D.S. & PEARSON R. (1963) Studies on the fungi of the root region. I. The development of fungi on young roots. *Pl. Soil* 19, 332–349

PEARMAN G.I. (1966) The reflection of visible radiation from leaves of some Western Australian species. *Aust. J. Biol. Sci.* **19**, 97–103

PEARS N.V. (1968) The natural altitudinal limit of forest in the Scottish Grampians. *Oikos* **19**, 71–80

PEARSON V. & READ D.J. (1973) The biology of mycorrhiza in the Ericaceae. II. The transport of carbon and phosphorus by the endophyte and the mycorrhiza. *New Phytol.* **72**, 1325–1331

PEEL A.J. (1965) On the conductivity of xylem in trees. *Ann. Bot.* **29**, 119–130

PENMAN H.L. (1948) Natural evaporation from open water, bare soil and grass. *Proc. R. Soc.* **A193**, 120–145

PENMAN H.L. & SCHOFIELD R.K. (1951) Some physical aspects of assimilation and transpiration. *Symp. Soc. Exp. Biol.* **5**, 115–129

PENNDORF R. (1956) Luminous reflectance (visual albedo) of natural objects. *Bull. Am. Met. Soc.* **37**, 142–144

PERRIER E.R., McKELL C.M. & DAVIDSON J.M. (1961) Plant-soil-water relationships in two subspecies of orchard grass. *Soil Sci.* **92**, 413–420

PFEFFER W. (1877) Über fleischfressende Pflanzen, über die Ernährung durch Aufnahme organischer Stoffe uberhaupt. *Landw. Jbr.* **6**, 969–998

PHILLIPS P.J. & McWILLIAM J.R. (1971) Thermal responses of primary carboxylating enzymes from C_3 and C_4 plants adapted to contrasting temperature environments. In: *Photosynthesis and photorespiration* (eds. M.D. Hatch, C.B. Osmond & R.O. Slatyer) pp. 97–104. Wiley-Interscience, New York

PIGOTT C.D. (1969) Influence of mineral nutrition on the zonation of flowering plants in coastal salt marshes. In: *Ecological aspects of the mineral nutrition of plants* (ed. I.H. Rorison) pp. 25–35. Blackwell Scientific Publications, Oxford

PIGOTT C.D. & TAYLOR K. (1964) The distribution of some woodland herbs in relation to the supply of nitrogen and phosphorus in the soil. *J. Ecol.* **52**, (supplement), 175–185

PISEK A. (1960) The nature of temperature optimum and minimum of photosynthesis. *Bull. Res. Counc. Israel* **8D**, 285–289

PISEK A. & BERGER E. (1938) Kutikuläre Transpiration und Trockenresistenz isolierter Blätter und Sprosse. *Planta* **28**, 124–155

PISEK A. & WINKLER E. (1953) Die Schliessbewegung der Stomata bei ökologische verschiedenen Pflanzentypen in Abhängkeit vom Wassersättigungszustand der Blätter und vom Licht. *Planta* **42**, 253–278

PISEK A. & WINKLER E. (1956) Wassersättigungsdefizit, Spaltenbewegung und Photosynthese. *Protoplasma* **46**, 597–611

PISEK A., LARCHER W. & UNTERHOLZNER R. (1967) Kardinaler Temperaturbereiche der Photosynthese und Grenztemperaturen des Lebens der Blätter verschiedener Spermatophyten. I. Temperaturminimum der Nettoassimilation, Gefrier- und Frostschadensbereiche der Blätter. *Flora, Jena* **157**, 239–264

PISEK A. & KEMNITZER R. (1968) Der Einfluss von Frost auf die Photosynthese der Weisstanne (*Abies alba* Mill). *Flora, Jena* **158**, 314–376

PISEK A., LARCHER W., PACK I. & UNTERHOLZNER R. (1968) Kardinale Temperaturbereiche der Photosynthese und Grenztemperaturen des Lebens der Blätter verschiedener Spermatophyten. II. Temperaturmaximum der Netto-Photosynthese und Hitzeresistenz der Blätter. *Flora, Jena* **158**, 110–128

PISEK A., LARCHER W., MOSER W. & PACK I. (1969) Kardinale Temperaturbereiche der Photosynthese und Grenztemperaturen des Lebens der Blätter
verschiedener Spermatophyten. III. Temperaturabhängigkeit und optimaler
Temperaturbereich der Netto-Photosynthese. *Flora, Jena* **158**, 608–630

POLWART A. (1970) Ecological aspects of the resistance of plants to environmental
factors. Ph.D. Thesis, University of Glasgow

POORE M.D. & MCVEAN D.N. (1956) A new approach to Scottish Mountain
vegetation. *J. Ecol.* **45**, 401–439

PREECE T.F. & DICKINSON C.H. (1971) *Ecology of leaf surface micro-organisms.*
Academic Press, London & New York

PROCTOR J. (1971a,b) The plant ecology of serpentine.
  II. Plant response to serpentine soils. *J. Ecol.* **59**, 397–410
  III. The influence of a high calcium/magnesium ratio and high nickel and
  chromium levels in some British and Swedish serpentine soils. *J. Ecol.*
  59, 827–842

PROCTOR M.C.F. & YEO P.(1973) *The pollination of flowers.* Collins,
London

PUTWAIN P.D. & HARPER J.L. (1970) Studies on the dynamics of plant populations. III. The influence of associated species on populations of *Rumex
acetosa* L. And *Rumex acetosella* L. in grassland. *J. Ecol.* **58**, 251–264

RACKHAM O. (1966) Radiation, transpiration and growth in a woodland annual.
In: *Light as an ecological factor* (eds. R. Bainbridge, G.C. Evans & O.
Rackham) pp. 167–185. Blackwell Scientific Publications, Oxford

RAINS D.W. & EPSTEIN E. (1967) Preferential absorbtion of potassium by leaf
tissue of mangrove; *Avicennia marina*. An aspect of halophytic competence in
coping with salt. *Aust J. Biol. Sci.* **20**, 847–857

RAMSAY J.A. & BROWN R.H.J. (1955). Simplified apparatus and procedure for
freezing point determination upon small volumes of fluid. *J. scient. Instrum.*
32, 372–375

RAMAKRISHNAN P.S. (1965) Studies on edaphic ecotypes in *Euphorbia thymifolia*
L. II. Growth performance, mineral uptake and interecotypic competition.
J. Ecol. **53**, 705–714

RAMAKRISHNAN P.S. (1968a,b, 1970) Nutritional requirements of the edaphic
ecotypes in *Melilotus alba* Medic.
  I. pH, calcium and phosphorus. *New Phytol.* **67**, 145–157
  II. Aluminium and manganese. *New Phytol.* **67**, 301–308
  III. Interference between calcareous and acid population in the two soil types.
  New Phytol. **69**, 81–86

RANFT H. & DÄSSLER H.G. (1970) Rauchhärtetest an Gehölzen im SO_2-Kabinen-
versuch. *Flora, Jena* **159**, 573–588

RASCHKE K. (1956) Über die physikalischen Beziehungen zwischen Wärmeüber-
gangszahl, Strahlungsaustausch Temperatur und Transpiration eines Blattes.
Planta **48**, 200–238

RASCHKE K. (1975) Stomatal action. *Ann. Rev. Pl. Physiol.* **26**, 309–340

REPP, G. (1963) Die Kupferresistez des Protoplasmas höheren Pflanzen auf
Kupfererzboden. *Protoplasma* **57**, 643–659

RICHARDS L.A. & OGATA G. (1958) Thermocouple for vapour pressure measurement in biological and soil systems at high humidity. *Science, N.Y.* **128**, 1089–1090

RIEDMÜLLER-SCHÖLM H.E. (1974) The temperature resistance of Alaskan plants from the Continental Boreal Zone. *Flora, Jena* **163**, 230–250

RITCHIE J.C. (1955) Biological Flora of the British Isles. *Vaccinium vitis-idaea* L. *J. Ecol.* **43**, 701–708

ROBERTS B.R. (1964) Effects of water stress on the translocation of photosynthetically assimilated carbon-14 in yellow poplar. In: *The function of wood in trees* (ed. M. Zimmerman) pp. 273–288. Academic Press, New York

ROBINSON R.K. (1971) The importance of soil toxicity in relation to the stabilisation of plant communities. In: *The scientific management of animal and plant communities for conservation* (eds. E. Duffey & A.S. Watt) pp. 105–113. Blackwell Scientific Publications, Oxford

ROBINSON R.K. (1972) The production by roots of *Calluna vulgaris* of a factor inhibitory to growth of some mycorrhizal fungi. *J. Ecol.* **60**, 219–224

ROBSON M.J. & JEWISS O.R. (1968) A comparison of British and North African varieties of tall fescue (*Festuca arundinacea*). II. Growth during winter and survival at low temperatures. *J. appl. Ecol.* **5**, 179–190

ROOK D.A. (1969) The influence of growing temperature on photosynthesis and respiration of Pinus radiata seedlings. *N.Z. Jl. Bot.* **7**, 43–55

RORISON I.H. (1967) A seedling bioassay of some soils in the Sheffield area. *J. Ecol.* **55**, 725–752

RORISON I.H. (1969) Ecological inferences from laboratory experiments in mineral nutrition. In: *Ecological aspects of the mineral nutrition of plants* (ed. I.H. Rorison) pp. 155–175. Blackwell Scientific Publications, Oxford

ROUATT J.W., KATZNELSON H. & PAYNE T.M.B. (1960) Statistical evaluation of the rhizosphere effect. *Proc. Soil Sci. Soc. Am.* **24**, 271–273

ROVIRA A.D. & BOWEN G.D. (1966) Phosphate incorporation by sterile and non-sterile plant roots. *Aust J. Biol. Sci.* **19**, 1167–1169

RUSSELL E.W. (1974) *Soil conditions and plant growth* (12th edn), Longmans, London

RUSSELL R.S. (1940) Physiological studies on an arctic vegetation. III. Observations on carbon assimilation, carbohydrate storage and stomatal movement in relation to the growth of plants on Jan Mayen Island. *J. Ecol.* **28**, 289–309

RUTTER A.J. (1955) Composition of wet heath vegetation in relation to the water table. *J. Ecol.* **43**, 507–543

RUTTER A.J. (1968) Water consumption by forests. In: *Water deficits and plant growth* Vol. II (ed. T.T. Kozlowski) pp. 23–84. Academic Press, New York

RYCHNOVSKÁ M. (1963) An outpost site of *Corynephorous canescens* in the region between the Danube and the Tisza and its causal explanation. *Acta Biol. Hung.* **14**, 57–66

RYCHNOVSKÁ M. (1965) Water relations of some steppe plants investigated by means of the reversibility of the water saturation deficit. In: *Water stress in plants* (ed. B. Slavík) pp. 108–116. Czech. Acad. Sci. Prague

RYCHNOVSKÁ M. & KVĚT J. (1963) Water relations of some psammophytes with respect to their distribution. In: *The water relations of plants* (eds. A.J. Rutter & F.H. Whitehead) pp. 190–198. Blackwell Scientific Publications, Oxford

SACHS J. (1860) Physiologische Untersuchungen über die Abhängigkeit der Keimung von der Temperatur. *Jb. wiss. Bot.* 2, 338–377

SALISBURY E.J. (1942) *The reproductive capacity of plants.* London

SANDS K. & RUTTER A.J. (1958) The relationship of leaf water deficit to soil moisture tension in *Pinus sylvestris* L. II. Variations in the relation caused by developmental and environmental factors. *New Phytol.* 57, 387–399

SAPPER I. (1935) Versuche zur Hitzeresistenze der Pflanzen. *Planta* 23, 518–556

SCHIMPER A.F.W. (1898) Pflanzengeographie auf physiol ogischer Grundlage. Fischer, Jena

SCHIMPER A.F.W. (1902) *Plant geography on a physiological basis.* Clarendon Press, Oxford

SCHMIDT A. (1891) Theoretische Verwertung der Königsberger Bodenttemperaturbeobachtungen. *Schr. phys.-ökon. Ges. Königsb.* 32, 97–168

SCHOFIELD R.K. (1935) The pF of water in soil. *Trans. 3rd. Int. Congr. Soil Sci.* 2, 37

SCHOLANDER P.F., HAMMEL H.T., HEMMINGSEN E. & CAREY W. (1962) Salt balance in mangroves. *Pl. Physiol., Lancaster* 37, 722–729

SCHOLANDER P.F., HAMMEL H.T., BRADSTREET E.D. & HEMMINGSEN E.A. (1965) Sap pressure in vascular plants. *Science, N.Y.* 48, 339–346

SCHRAMM J.R. (1966) Plant colonization studies on black wastes from anthracite mining in Pennsylvania. *Trans. amer. phil. Soc.* 56, 1–194

SCHRATZ E. (1932) Untersuchung über die Beziehung zwischen Transpiration und Blattstruktur. *Planta* 16, 17–69

SCHROEDER D. (1969) *Bodenkunde in Stichworten* F. Hirt, Kiel

SCHULZE E.D. (1970) Der CO_2-Gaswechsel der Buche (*Fagus silvatica* L.) in Abhängigkeit von den Klimafaktoren im Freiland. *Flora, Jena* 159, 177–232

SCHWARZ W. (1970) Der Einfluss der Photoperiode auf der Austreiben, die Frosthärte und die Hitzeresistenz von Zirben un Alpenrosen. *Flora, Jena* 159, 258–285

SCOTT E.A. (1971) The resistance of *Calluna vulgaris* to herbicide. B.A. (Hons) Thesis. University of Stirling

ŠESTÁK Z., ČATSKÝ J. & JARVIS P.G. (1971) *Plant photosynthetic production. Manual of Methods.* Dr. W. Junk N.V., The Hague

SHAH C.B. & LOOMIS R.S. (1965) Ribonucleic acid and protein metabolism in sugar beet during drought. *Physiol Pl.* 18, 240–254

SHONTZ N.N. & SHONTZ J.P. (1972) Competition for nutrients between ecotypes of *Galinsoga ciliata. J. Ecol.* 60, 89–92

SLATYER R.O. (1960) Internal water blance of *Acacia aneura* F. Muell. in relation to environmental conditions. *Arid. Zone Res.* 16, 137–146

SLATYER R.O. (1967) *Plant-water relationships.* Academic Press, New York

SMITH S.E. (1966) Physiology and ecology of *Orchis* mycorrhizal fungi with reference to seedling nutrition. *New Phytol.* 65, 488–499

SMITH D., MUSCATINE L. & LEWIS D.H. (1969) Carbohydrate movement from autotrophs to heterotrophs in parasitic and mutualistic symbiosis. *Biol. Rev.*, **44**, 17–85

SMITH K.A. & RESTALL S.W.F. (1971) The occurrence of ethylene in anaerobic soil. *J. Soil Sci.* **22**, 430–443

SMITHBERG M.H. & WEISER C.J. (1968) Pattern of variation among climatic races of red osier dogwood. *Ecology* **49**, 495–505

SNAYDON R.W. (1962) Micro-distribution of *Trifolium repens* and its relation to soil factors. *J. Ecol.* **50**, 133–143

SPANNER D.C. (1951) The Peltier Effect and its use in the measurement of suction pressure. *J. exp. Bot.* **2**, 145–168

SPARLING J.H. (1967) The occurrence of *Schoenus nigricans* L. in blanket bogs. II. Experiments on the growth of *Schoenus nigricans* under controlled conditions. *J. Ecol.* **55**, 15–31

SPARLING J.H. (1968) Biological Flora of the British Isles. *Schoenus nigricans* L. *J. Ecol.* **56**, 883–899

STEVENSON A.G. (1972) Interference by *Mercurialis perennis* on other species. B.A.(Hons) Thesis, University of Stirling

STEWART W.D.P. & PEARSON M.C. (1967) Nodulation and nitrogen fixation by *Hippophaë rhamnoides* in the field. *Pl. Soil* **26**, 348–360

STEWART W.D.P. (1974) Blue green algae. In: *The biology of nitrogen fixation* (ed. A. Quispel) pp. 202–237. North Holland Publishing Co., Amsterdam and Oxford

STEWART W.S. & BANNISTER P. (1973) Seasonal changes in carbohydrate content of three *Vaccinium* spp. with particular reference to *V. uliginosum* L. and its distribution in the British Isles. *Flora, Jena* **162**, 134–155

STEWART W.S. & BANNISTER P. (1974) Dark respiration rates in *Vaccinium* spp. in relation to altitude. *Flora, Jena* **163**, 415–421

STEWART V.I. & ADAMS W.A. (1968) The quantitative description of soil moisture states in natural habitats with special reference to moist soils. In: *The measurement of environmental factors in terrestrial ecology* (ed. R.M. Wadsworth) pp. 161–173. Blackwell Scientific Publications, Oxford

STOCKER O. (1929) Die Wasserdefizit von Gefässpflanzen in verschiedenen Klimazonen. *Planta* **7**, 382–387

STOCKER O. (1956) Die Abhängigkeit der Transpiration von den Umvweltfaktoren. In: *Handbuch der Pflanzenphysiologie* (ed. W. Ruhland) pp. 436–488. Springer, Berlin

STONE J.F., KIRKHAM D. & READ A.A. (1955) Soil moisture determination by a portable neutron scatter meter. *Proc. Soil. Soc. Am.* **19**, 418–423

STREETER D.T. (1971) Public pressure on the vegetation of chalk downland at Box Hill, Surrey. In: *The scientific management of animal and plant communities for conservation* (eds. E. Duffey & A.S. Watt) pp. 459–468. Blackwell Scientific Publications, Oxford

STRIBLEY D.P. & READ D.J. (1974a,b) The biology of mycorrhiza in the Ericaceae. III. Movement of carbon-14 from host to fungus. *New Phytol.* **73**, 731–741
IV. The effect of mycorrhizal infection on the uptake of ^{15}N from labelled soil by *Vaccinium macrocarpon* Ait. *New Phytol.* **73**, 1149–1155

SUTTON C.D. & GUNARY D. (1969) Phosphate equilibria in soil. In: *Ecological aspects of the mineral nutrition of plants* (ed. I.H. Rorison) pp. 127–134. Blackwell Scientific Publications, Oxford

SWEENEY J.R. (1956) Responses of vegetation to fire. *Univ. Calif. Publ. Bot.* **28**, 143–250

SZAREK S.R. & TING I.P. (1974) Seasonal patterns of acid metabolism and gas exchange in *Opuntia basiliaris*. *Pl. Physiol.*, *Lancaster* **54**, 76–81

SZEICZ G. (1966) Field measurements of energy in the 0·4–0·7 micron range. In: *Light as an ecological factor* (eds. R. Bainbridge, G.C. Evans & O. Rackham) pp. 41–52. Blackwell Scientific Publications, Oxford

TANSLEY A.G. (1939) *The British Isles and their vegetation*. Cambridge University Press, Cambridge

TAYLOR K. (1971) Biological flora of the British Isles. *Rubus chamaemorus* L. *J. Ecol.* **59**, 293–306

TAYLOR S.A. & SLATYER R.O. (1961) Proposals for a unified terminology in studies of plant-soil water relationships. In: *Plant-water relationships in arid and semi-arid conditions* (Proc. Madrid Symp.) pp. 339–349. UNESCO, Paris

TEICHERT F. (1955) Vergleichende Messungen des Ozonegehaltes der Luft an Erdboden und im 80m Höhe. *Z. Met.* **21**, (7), 21–27

THOM A.S. (1968) The exchange of momentum, mass and heat between an artificial leaf and the airflow in a wind tunnel. *Q. Jl R. met. Soc.* **94**, 44–55

THOMPSON P.A. (1969) Comparative effects of gibberellins A₃ and A₄ on the germination of seeds of several different species. *Hort. Res.* **9**, 130–138

THOMPSON P.A. (1970a) Germination of species of Caryophyllaceae in relation to their geographical distribution in Europe. *Ann. Bot.* **34**, 427–449

THOMPSON P.A. (1970b) A comparison of the germination character of species of Caryophyllaceae collected in Central Germany. *J. Ecol.* **58**, 699–711

THORP T.K. (1972) The ecology and phytosociology of heathland in south-west Scotland. Ph.D. Thesis, University of Glasgow

THURMANN J. (1849) Essai du phytostatique appliqué à la chaine du Jura. Berne

THURSTON J.M. (1969) The effect of liming and fertilisers on the botanical composition of permanent grassland and on the yield of hay. In: *Ecological aspects of the mineral nutrition of plants* (ed. I.H. Rorison) pp. 3–10. Blackwell Scientific Publications. Oxford

TILL O. (1956) Über die Frosthärte von Pflanzen sommergrüner Laubwald. *Flora, Jena* **143**, 499–542

TIMSON J. (1966) Biological flora of the British Isles. *Polygonum hydropiper* L. *J. Ecol.* **54**, 815–821

TING I.P. (1971) Nonautotrophic CO₂ fixation and crassulacean acid metabolism. In: *Photosynthesis and photorespiration* (eds. M.D. Hatch, C.B. Osmond & R.O. Slatyer) pp. 169–185. Wiley-Interscience, New York

TINKER P.B. (1969) The transport of ions in the soil around plant roots. In: *Ecological aspects of the mineral nutrition of plants* (ed. I.H. Rorison) pp. 135–147. Blackwell Scientific Publications, Oxford

TINKLIN R. & WEATHERLEY P.E. (1966) The role of root resistance in the control of leaf water potential. *New Phytol.* **65**, 509–517

TINKLIN R. & WEATHERLEY P.E. (1968) The effect of transpiration rate on the leaf water potential of sand and soil rooted plants. *New Phytol.* **67**, 605–615

TOLBERT N.E. (1971) Leaf peroxisomes and photorespiration. In: *Photosynthesis and photorespiration* (eds. M.D. Hatch, C.B. Osmond & R.O. Slatyer) pp. 458–471. Wiley-Interscience, New York

TRANQUILLINI W. (1970) Einfluss des Windes auf den Gaswechsel der Pflanzen. *Umschau*, **26**, 860–861

TRESHOW M. (1970) *Environment and pant responses.* McGraw-Hill, New York

TRUOG E. (1947) Soil reaction influence on availability of plant nutrients. *Proc. Soil Soc. Am.* **11**, 305–308

TURNER N.C. (1970) Response of adaxial and abaxial stomata to light. *New Phytol.* **69**, 647–653

TURNER R.G. (1969) Heavy metal tolerance in plants. In: *Ecological aspects of the mineral nutrition of plants* (ed. I.H. Rorison) pp. 339–410. Blackwell Scientific Publications, Oxford

TYLER P.D. & CRAWFORD R.M.M. (1970) The role of shikimic acid in waterlogged roots and rhizomes of *Iris pseudacorus* L. *J. exp. Bot.* **21**, 677–682

ULMER W. (1937) Über den Jahresgang der Frosthärte einiger imergrüner Arten der alpinen Stüfe, sowie der Zirbe und Fichte. *Jb. wiss. Bot.* **84**, 552–592

UNGEHEURER H. (1934) Mikroklima in einem Buchenhochwald am Hang. *Bioklim. Beibl.* **1**, 75–88

UNGAR I.A. & HOGAN W.C. (1970) Seed germination in *Iva annua*. *Ecology*, **51**, 150–154

UNGER FR. (1836) Über den Einfluss des Bodens auf die Verteilung der Gewächse, nachgewiesen in der Vegetation des nordöstlichen Tirols. Vienna

UNGERSON J. & SCHERDIN G. (1968) Jahresgang von Photosynthese und Atmung unter natürlichen Bedingungen bei *Pinus sylvestris* L. an ihrer Nordgrenze in der Subarktis. *Flora, Jena* **157**, 391–434

URSPRUNG A. & BLUM G. (1916) Zur Kenntnis der Saugkraft. *Ber dt. bot. Ges.* **34**, 525–539

VÄUPEL A. (1958) Mikroklima und Pflanzentemperaturen auf trocken-heissen Standorten. *Flora, Jena* **145**, 497–541

WALLACE T. (1961) *The diagnosis of mineral deficiencies in plants by visual symptoms.* HMSO, London

WALLEY K.A., KHAN M.S.I. & BRADSHAW A.D. (1974) The potential for evolution of heavy metal tolerance in plants. I. Copper and zinc tolerance in *Agrostis tenuis*. *Heredity, Lond.* **32**, 309–319

WALTER H. (1931) *Die Hydratur der Pflanzen*, Fischer, Jena

WALTER H. (1960) *Einführung in die Phytologie. III/I Standortslehre.* (2nd edn) Ulmer, Stuttgart

WALTER H. (1963) Zur Klärung des spezifischen Wasserzustandes im Plasma. II. Methodisches *Ber. dt. bot. Ges.* **76**, 54–71

WARREN-WILSON J. (1957) Arctic plant growth. *Advmt. Sci., Lond.* **13**, 383–388

WATERHOUSE F.L. (1955) Microclimatological profiles in grass cover in relation to biological problems. *Q. Jl. R. met. Soc.* **81**, 63–71

WARMING E. (1896) *Plantesamfund.* Copenhagen.

WARMING E. (1909) *Ecology of plants.* Clarendon Press, Oxford (English translation)

WEATHERLEY P.E. (1950) Studies in the water relations of the cotton plant. I. The field measurement of water deficits in leaves. *New Phytol.* **48**, 81–97

WEATHERLEY P.E. & SLATYER R.O. (1957) The relationship between relative turgidity and diffusion pressure deficit in leaves. *Nature, Lond.* **179**, 1085–1086

WEBLEY D.M., EASTWOOD D.J. & GIMINGHAM C.H. (1952) Development of soil microflora in relation to plant succession including rhizosphere flora associated with colonising species. *J. Ecol.* **40**, 168–178

WEBSTER J.R. (1962a,b) The composition of wet heath vegetation in relation to aeration of the groundwater and soil.

I. Field studies of groundwater and soil aeration in several communities· *J. Ecol.* **50**, 619–637

II. Response of *Molinia caerulea* to controlled conditions of soil aeration and ground-water movement. *J. Ecol.* **50**, 639–650

WEISSENBÖCK G. (1969) Der Einfluss des Bodensalzgehaltes auf Morphölogie und Ionenspeicherung von Halophyten. *Flora, Jena* **158**, 369–389

WENT F.W. (1970) Plants and the chemical environment. In: *Chemical ecology* (eds. E. Sondheimer & J.B. Simeone) pp. 71–82. Academic Press, New York

WESSON G. & WAREING P.F. (1967) Light requirements of buried seeds. *Nature, Lond.* **213**, 600–601

WHITEHEAD F.H. (1963) The effects of exposure on growth and development. In: *The water relations of plants* (eds. A.J. Rutter & F.H. Whitehead) pp. 235–245. Blackwell Scientific Publications, Oxford

WHITMORE T.C. & WONG Y.K. (1959) Patterns of sunfleck and shade light in tropical rain forest. *Malay, Forester* **22**, 50–62

WHITTAKER E. & GIMINGHAM C.H. (1962) The effects of fire on the regeneration of *Calluna vulgaris* (L.) Hull from seed. *J. Ecol.* **50**, 815–822

WHITTAKER R.H. (1970a) The biochemical ecology of higher plants. In: *Chemical ecology* (eds. E. Sondheimer & J.B. Simeone) pp. 43–70. Academic Press, New York

WHITTAKER R.H. (1970b) *Communities and Ecosystems.* Macmillan, New York

WHITWORTH J.W. & MUSIK T.J. (1967) Differential response of selected clones of bindweed to 2,4-D. *Weeds* **15**, 275–280

WIESNER J. (1907) *Der Lichtgenuss der Pflanzen.* Fischer, Leipzig

WILLIAMS C.M. (1970) Hormonal interactions between plants and insects. In: *Chemical ecology* (eds. E. Sondheimer & J.B. Simeone) pp. 103–132. Academic Press, New York

WILLIS A.J. (1963) Braunton Burrows: the effects on the vegetation of the addition of mineral nutrients to the dune soils. *J. Ecol.* **51**, 353–374

WILLIS A.J. (1965) The influence of mineral nutrients on the growth of *Ammophila arenaria. J. Ecol.* **53**, 735–745

WIND G.P. (1955) Flow of water through plants. *Neth. J. Agric. Sci.* **3**, 259–264

WIT C.T. DE (1960) On competition. *Versl. lanbouwk onderz. Ned.* **66** (8), 1–82

WIT C.T. DE, BERGH J.P. VAN DEN (1965) Competition between herbage plants. *Neth. J. agric. Sci.* **13**, 212–221

Woods D.B. & Turner N.C. (1971) Stomatal response to changing light by four tree species of varying shade tolerance. *New Phytol.* **70**, 77–84

Woolhouse H.W. (1969) Differences in the properties of the acid phosphatases of plant roots and their significance in the evolution of edaphic ecotypes. In: *Ecological aspects of the mineral nutrition of plants* (ed. I.H. Rorison) pp. 357–380. Blackwell Scientific Publications, Oxford

Wu L. & Antonovics J. (1975) Zinc and copper uptake by *Agrostis stolonifera*, tolerant to both zinc and copper. *New Phytol.* **75**, 231–237

Wurster C.F. (1968) D.D.T. reduces photosynthesis in marine phytoplankton. *Science, N.Y.* **159**, 1474–1475

Wyn-Jones R.G. & Lunt O.R. (1967) Function of calcium in plants. *Bot. Rev.* **33**, 407–426

Zenker H. (1954) Waldeinfluss auf Kondensationskerne und Lufthygiene. *Z. Met.* **8**, 150–159

Ziegler I. (1972). The effects of $So_3{}^{--}$ on the activity of ribulose-1,5 disphosphate carboxylase in isolated spinach chloroplasts. *Planta* **103**, 155–163

REFERENCES ADDED IN PROOF

Coombe D. E. (1966) The seasonal light climate and plant growth in a Cambridgeshire wocd. In: *Light as an Ecological Factor* (ed. R. Bainbridge, G. C. Evans & O. Rackham) pp. 148–166. Blackwell Scientific Publications, Oxford

Effenberger, E. F. (1940) Kern- und Staubuntersuchungen am Collmberg. *Veröff. geophys. Inst. Univ. Leipzig.* **12**, 305–359

Henning H. (1957) Pico-aerologische Untersuchungen über Temperatur- und Windverhältnisse der bodennahen Luftschicht bis 10m Höhe in Lindenberg. *Abh. Met. hydrol. Dienst. D.D.R.* **6**, 1–166

Huber B. (1952) Über die vertikale Reichwette vegetationsbedingter Tagesschwankungen im CO_2-Gehalt der Atmosphäre. *Forstwiss Cblt.* **71**, 372–380

Kramer P. J. (1951) Causes of injury to plants resulting from flooding of the soil. *Pl. Physiol., Lancaster,* **26**, 722–36

Meetham A. R. (1956) *Atmospheric pollution.* Pergamon Press, London

Milthorpe F. L. (1961) Plant factors involved in transpiration. In: *Plant-water relationships in arid and semi-arid conditions.* pp. 107–113, UNESCO, Paris

Author Index

Italicized numbers indicate inclusion in tables, figures or references. Only authors mentioned in the text are included; consequently some second and third authors may not be mentioned.

Species Index

Includes species and generalized references to groups of plants. Italicized numbers indicate inclusion in tables, figures or references.

264 INDEX

Subject Index